Springer Series in Language and Communication 14

Editor: W. J. M. Levelt

Springer Series in Language and Communication

Editor: W. J. M. Levelt

Volume 1 **Developing Grammars**
By W. Klein and N. Dittmar

Volume 2 **The Child's Conception of Language** 2nd Printing
Editors: A. Sinclair, R. J. Jarvella, and W. J. M. Levelt

Volume 3 **The Logic of Language Development in Early Childhood**
By M. Miller

Volume 4 **Inferring from Language** By L. G. M. Noordman

Volume 5 **Retrieval from Semantic Memory**
By W. Noordman-Vonk

Volume 6 **Semantics from Different Points of View**
Editors: R. Bäuerle, U. Egli, A. von Stechow

Volume 7 **Lectures on Language Performance** By Ch. E. Osgood

Volume 8 **Speech Act Classification**
By Th. Ballmer and W. Brennenstuhl

Volume 9 **The Development of Metalinguistic Abilities in Children**
By D. T. Hakes

Volume 10 **Modelling Language Behaviour**
By R. Narasimhan

Volume 11 **Language in Primates: Perspectives and Implications**
Editors: J. de Luce and H. T. Wilder

Volume 12 **Concept Development and the Development of Word Meaning**
Editors: Th. B. Seiler and W. Wannenmacher

Volume 13 **The Sun is Feminine**
A Study on Language Acquisition in Bilingual Children
By T. Taeschner

Volume 14 **Prosody: Models and Measurements**
Editors: A. Cutler and D. R. Ladd

Volume 15 **Language Awareness in Children** By D. L. Herriman

Volume 16 **Dynamic Aspects of Language Processing: Focus and Presupposition**
By J. Engelkamp and H. D. Zimmer

Prosody: Models and Measurements

Edited by A. Cutler and D. R. Ladd

With Contributions by
G. Brown H. Buxton A. Cutler E. Gårding
D. Hirst D. R. Ladd M. Nespor N. Thorsen
J. Vaissière I. Vogel

With 32 Figures

Springer-Verlag
Berlin Heidelberg New York Tokyo 1983

Dr. Anne Cutler

MRC Applied Psychology Unit, 15 Chaucer Road,
Cambridge CB2 2EF, United Kingdom

Dr. D. Robert Ladd

Fachbereich Psychologie, Justus-Liebig-Universität Gießen,
Otto-Behaghel-Str. 10, 6300 Gießen, Fed. Rep. of Germany

Series Editor:

Professor Dr. Willem J. M. Levelt

Max-Planck-Institut für Psycholinguistik, Berg en Dalseweg 79,
6522 BC Nijmegen, The Netherlands

ISBN-13: 978-3-642-69105-8 e-ISBN-13: 978-3-642-69103-4
DOI: 10.1007/978-3-642-69103-4

Library of Congress Cataloging in Publication Data. Main entry under title: Prosody, models and measurements. (Springer series in language and communication ; v. 14) 1. Prosodic analysis (Linguistics)–Addresses, essays, lectures. I. Cutler, A. (Anne) II. Ladd, D. Robert, 1947- . III. Brown, Gillian. IV. Series. P224.P76 1983 414 83-4243

Softcover reprint of the hardcover 1st edition 1983

Typesetting: Schwetzinger Verlagsdruckerei GmbH

Preface

Prosody: Models and Measurements is the fruit of a three-day workshop held in Paris in April, 1982. The workshop was one of a series which is sponsored by the Maison des Sciences de l'Homme under the auspices of the European Psycholinguistics Association, and which aims to bring together workers in a particular field from different European laboratories and to encourage future collaboration across regional, national and disciplinary boundaries. Thus the topic of the workshop – "Prosody" – was fairly liberally interpreted in the invitations, and the participants were drawn from a variety of backgrounds – linguistics, phonetics, psychology. Despite this diversity, however, there was a surprising degree of congruence in the topics discussed and points of view adopted. We have attempted in the introduction to the present volume to draw out the similarities between approaches, as well as to account for the differences and to set the individual contributions against the background of current research in the field.

The book is not merely a transcript of what was said during the sessions. All of the papers have been rewritten by the participants – in some cases extensively – to reflect the comments that were made in discussion and the points of contact and disagreement that became evident during the three days. Because the diversity of backgrounds has inevitably led to considerable terminological variation, we have added a set of notes on the terms and topics which appear in the individual contributions, in order to make comparison easier. It is hoped that the result is a coherent collection that fairly mirrors the state of prosodic research in Europe today.

Although all the contributions in this volume were presented at the workshop, they do not constitute an exhaustive report: M. Bierwisch (Berlin) and S. G. Nooteboom (Eindhoven) were invited but were unable to attend, and W. Klein (Nijmegen) and D. Weniger (Zürich) presented papers at the workshop which do not appear in the volume.

To Dr. Clemens Heller and the Maison des Sciences de l'Homme, who sponsored the workshop, to Mme. Adriana Touraine of the M.S.H., who took care of all the administrative details, and to the executive committee of the European Psycholinguistics Association, who encouraged and supported us at all points, we would like to express our very deep gratitude. The editors' collaboration was greatly facilitated by the Max-Planck-Gesellschaft, which awarded the first editor a Visiting Scholarship at the Max-Planck-Institut für Psycholinguistik, Nijmegen, during the tenure of which work on this volume was completed.

Nijmegen and Giessen, October 1982 *A. Cutler · D. R. Ladd*

Contents

List of Contributors

Brown, Gillian
Department of Linguistics, University of Edinburgh, George Square, Edinburgh EH8 9LL, U.K.

Buxton, Hilary
Artificial Intelligence Group, Department of Computer Science, Queen Mary College, University of London, Mile End Rd., London E1, U.K.

Cutler, Anne
MRC Applied Psychology Unit, 15 Chaucer Rd., Cambridge CB2 2EF, U.K.

Gårding, Eva
Department of Linguistics and Phonetics, University of Lund, Helgonabacken 12, S-223 62 Lund, Sweden

Hirst, Daniel
CNRS Institut de phonétique, Université de Provence, 29 avenue Robert Schuman, F-13621 Aix, France

Ladd, D. Robert
Fachbereich Psychologie, Justus-Liebig-Universität Gießen, Otto-Behaghel-Str. 10, 6300 Gießen, Fed. Rep. of Germany

Nespor, Marina
Department of Linguistics, University of Amsterdam, Spuistraat 210, 1012 VT Amsterdam, The Netherlands

Thorsen, Nina
Institute of Phonetics, University of Copenhagen, Njalsgade 96, DK-2300 Copenhagen S, Denmark

Vaissière, Jacqueline
Centre National d'Etudes des Télécommunications, Route de Tregastel, F-22301 Lannion, France

Vogel, Irene
Institute of Romance Philology, University of Venice, San Sebastiano, Venezia, Italy

1. Introduction. **Models and Measurements in the Study of Prosody**

D. Robert Ladd and Anne Cutler

1.1 Concrete and Abstract Accounts of Prosody

There are two broad traditions in the study of prosody that may be characterized – or caricatured – by their methodological preferences for one or the other of the scientific activities mentioned in the title: making measurements and constructing models. On one side of the dichotomy stand instrumental and experimental studies that seek to quantify acoustic features and investigate perceptual responses. On the other are descriptive and theoretical studies of prosodic structure and its relation to other aspects of grammar and phonology. In a great deal of past work these two traditions have simply ignored one another.

Occasionally, however, the differences between the two traditions have surfaced in explicit condemnation of the model-builders by the measurers. A pointed example can be found in Ohala's review of Lehiste 1970, in which he criticizes Lehiste for "faithfully set[ting] forth the full repertory of clever solutions which linguists have been able to come up with .... The catalog of such facile inventions by taxonomic linguists ... seems out of place in a book otherwise devoted to what scientists of language have proven and demonstrated empirically about the behavior of speech sounds" [Ohala, 1975, pp. 737–738]. The implication of such criticisms is that the two traditions are based on fundamentally different assumptions about what constitutes empirical work.

We disagree. We believe that the measurers and the model-builders have fundamentally different assumptions not primarily about methodology, but about function and representation of prosody. These assumptions can be explicitly stated and carefully examined: the issues are potentially empirical ones that can be tested in ways that will satisfy a range of methodological preferences. In particular, it seems to us that two of the main points of substantive disagreement implicit in the differences of methodology have to do with the *abstractness* of prosodic representation and the *directness* of the link between prosodic function and prosodic form.

One approach – let us call it the "concrete" approach – defines prosody more or less in physical terms, as *those phenomena that involve the acoustic parameters of pitch, duration, and intensity.* This approach conceives of the link between form and function in prosody as a relatively direct mapping between concrete meanings or functions and specific acoustic shapes or variables. Prosodic cues convey a relatively heterogeneous range of specific meanings – from "anger" to "phrase boundary" to "emphasis" – and constitute a kind of parallel channel or accompaniment to the central linguistic message of the utterance. Notions such as "continuation rise", "ironic intonation", and even "contrastive stress" are theoretical constructs of this general point of view. The main aim of "concrete" research is to identify the correlations between specific messages and specific acoustic parameters.

Compared to the concrete approach, the "abstract" approach sees prosody more from the point of view of its place in linguistic structure than its phonetic nature, and tends to class as prosodic *any phenomena that involve phonological organization at levels above the segment.* In practice, of course, the two definitions cover much of the same ground – intonation, stress, phrasing, etc. – but the different basis of the two approaches can be seen by the fact that certain phenomena, such as vowel harmony, might be considered "prosodic" by the abstract definition but not by the concrete one, while the reverse is true for phonemic vowel length, for example.

The difference of definition entails clearly distinct assumptions about function. In particular, the abstract approach does not assume that the relation between form and function is so straightforward as to justify the search for direct correlates. Prosody is not a phonetically defined component of language with a set of heterogeneous functions like "phrase boundary" and "question"; such messages are not conveyed by the prosody as a peripheral channel running parallel to the text, but as part of the whole linguistic structure, of which the prosodic features are one aspect. Acoustic parameters are seen as cues to abstract formal categories, not to concrete functional ones, and the main aim of prosodic research is to model the formal system.

1.2 Illustrating the Two Approaches

In order to illustrate the interaction of theoretical assumptions with methodological preferences, let us consider examples of the concrete and abstract approaches to two specific topics in prosodic research: the function of intonation, and the nature of stress.

1.2.1 Intonation

A good deal of "concrete" research on the relation between intonation and grammar treats fundamental frequency and duration as the phonetic accompaniment of certain types of syntactic units or constituent boundaries, and measures the acoustic characteristics of prosodic features in different syntactic environments. The most ambitiously reductionist work in this tradition is undoubtedly that of Lieberman [1967], who claimed that all the linguistically relevant uses of intonation could be reduced to a single, innate contrast between "marked and unmarked breath group", plus a prominence feature of increased subglottal pressure. Less controversial work along these general lines concentrates on specific syntactic environments such as yes/no questions, non-terminal boundaries, parenthetical phrases, echo questions, and so on. A great deal of work on French has taken this approach [Delattre, 1963; Martin, 1980; Grundstrom and Léon, 1973], as has recent work whose emphasis is on developing prosodic data for use in speech synthesis [O'Shaughnessy, 1976; Klatt, 1979]. Similar assumptions are made in Lehiste's [1975] work on paragraph cues and in much of the recent work on "declination" and its relation to syntax [Collier, 1975b; Maeda, 1976; Cooper and Sorensen, 1981].

In the same way, a direct connection is often assumed to exist between the prosodic characteristics of speech and various affective messages. Research on the vocal expression of emotion often treats the entire non-segmental part of the speech signal as a kind of overlay directly communicating emotional state. Acoustic correlates of the emotions signalled in this manner are frequently analyzed in terms of parameters that include such things as average fundamental frequency and overall frequency range [Williams and Stevens, 1972; Scherer, 1981]. Note that this work, when taken together with the work cited in the preceding paragraph, implies the existence of potential conflicts or interactions between the grammatical and emotional specifications of prosody. While this has not, to our knowledge, been seriously investigated, it is generally assumed that on the one hand, the two types of effects are often manifested in different ways (e.g. in the distinction between intonation and voice quality), and on the other hand, that interactions between the two uses, in particular on fundamental frequency, will ultimately be specifiable in precise acoustic terms.

The relation between grammatical and affective functions of intonation is seen quite differently in a wide range of "abstract" research, in particular the work done by a long line of scholars of intonation whose native language is English. Even a superficial look at English intonation is enough to make one suspect that the apparently different grammatical and emotional functions may actually be expressed by the same set of formal distinctions: a contour that is neutral in one context may express marked affect in another. Consequently, much work on English intonation does not start from a basic set of intonational *functions,* but assumes instead the existence of an inventory of more abstract intonational *categories* (e.g. "nuclear tones"). This emphasis has meant that functional labels have been used mostly as unsystematic glosses of the linguistic distinctions under investigation; most older work on English (e.g. the British pedagogical tradition of Kingdon [1958], O'Connor and Arnold [1961], etc.) simply mixes grammatical and affective descriptions in discussing intonational meanings (e.g. peremptory request, surprised question).

However, such descriptions are consistent with more recent explicit attempts to explain the surface heterogeneity of intonational function in terms of some deeper unity, specifically those explanations which rely (in current terminology) on the notion of pragmatic inference. This approach was first suggested by Pike [1945], who assigned approximate abstract meanings to intonation contours and left the context to narrow the range of specific interpretations; Pike specifically argued against the use of concepts like "question intonation". Explanations in this vein have been advanced more recently by Bolinger [e.g. 1958, 1982], Liberman and Sag [1974], Menn [1976], Ladd [1980], Brazil et al. [1980], and Cruttenden [1981]. Note also that this abstract approach is not inconsistent with the more experimentally oriented work of 't Hart and Collier [e.g. 1975] or Pierrehumbert [1981], which assumes that any description of intonational meaning must be in terms of the abstract phonological units of the intonational analysis, not in terms of specific grammatical or affective functions.

1.2.2 Stress

Similar disagreements between the concrete and abstract approaches can be found in studies on stress: here the issue is not so much the directness of the link between form and function, as the abstractness of the linguistic categories under investigation. The observation to be accounted for is that one or more of the syllables in a word or utterance are somehow more prominent than the rest. A great deal of "concrete" research (starting with Fry [1955] and a number of others in the 1950s and continuing through such recent work as Gay [1978]) has shown clearly that increases or changes in pitch and increases in intensity and duration on a given syllable are highly correlated with judgments that that syllable is the most prominent in a word or utterance. A great deal of "abstract" research, meanwhile (going back to Newman [1946] and continuing through Chomsky and Halle [1968] to the recent work inspired by Liberman and Prince [1977]), has concentrated instead on the patterns of prominence in lexical derivation, and on relative prominence of syllables other than the most stressed. In the first view, stress is a relatively concrete feature realized by specific acoustic cues; a syllable either has it or does not have it. In the second view, stress is a relatively abstract feature that specifies the relative prominence of syllables in a prosodic structure; the whole structure – not individual syllable features – determines the acoustic details of pitch, duration, and so on.

1.3 Implications of the Two Approaches

By now it should be clear how one's approach to investigating prosody reflects not only methodological preference, but also the very conception of the questions to be investigated. The model-builder is interested in establishing an inventory of abstract categories – a formal *representation* – of prosodic function and prosodic form. The goal of the model-builder's enterprise is to describe the systematic structure underlying prosodic distinctions; the basic assumption is that there are well-defined abstract levels of representation that mediate between specific prosodic functions like "phrase boundary" and specific acoustic traits.

By contrast, for many of those who take the measurer's approach, the primary concern is not representation, but *realization*. The question being asked is: What are the physical correlates of this or that prosodic message? To the extent that such investigators have constructed explicit models of prosodic representation, they have tended to think in terms not of linguistic categories, but of interacting parameters; their models assign acoustic correlates to individual functions such as word stress, sentence stress, sentence modality, affective use of pitch range, and so on, and attempt to specify the interaction of all these effects on individual parameters like fundamental frequency.

When the two positions are expressed in this way, it becomes clear that one of the main issues between them is what we might call "the place of prosody in language". Some of the functions normally considered under the rubric of prosody do seem to represent a separate communicative channel that accompanies the verbal message but is peripheral to it: the least controversial example is perhaps the differences of voice quality that accompany different emotional states. The disagreement arises over the extent to which this peripheral-channel model can properly be applied to other prosodic phenomena. Descriptions of prosody that contain abstract phonological categories or assign a central role to pragmatic inference take notions that have proven useful in the description of other aspects of language and apply them to prosody; this implies a belief that in at least certain respects prosody is an integral part of language. The experimental view, insofar as it assumes direct links between physiology and prosodic function, or describes prosodic form in terms of interacting parametric variables rather than abstract categories, treats prosody more as a separate channel than as part of the basic structure itself.

The issue of whether prosody is "central" or "peripheral" is surely not a simple empirical question with an either/or answer, but it does suggest more specific empirical questions and hypotheses that could reflect on the validity of the assumptions that underlie the different approaches to prosody. For example, clinical data and neurolinguistic research are a valuable source of empirical evidence whose implications are seldom taken into account. Such evidence suggests that in some ways prosodic phenomena are indeed peripheral to normal language function: for example, it is well known that even global aphasics can often express emotion vocally, and there is some experimental evidence [Blumstein and Cooper, 1974; Van Lancker and Fromkin, 1973] to suggest that intonational cues are processed differently from other linguistic stimuli. Other experimental evidence [Zurif and Mendelsohn, 1972] casts doubt on the view that prosody is processed separately, and other clinical data suggest that certain prosodic distinctions may be lost in aphasia with the rest of language (for a review see Weniger [1978]). An important contribution to resolving the issue of prosody's role in language could be made by experiments and clinical studies designed to test specific hypotheses.

In an introduction of this sort we cannot properly take sides on the general issue of concrete and abstract models of prosodic function and representation, since the contributors whose work we are introducing clearly represent different points of view. We have, however, tried here to place the papers in a context that emphasizes the interdependence – even the inseparability – of making measurements and constructing models. Those whose methodological preference runs to experimental work must acknowledge that experimental data are not theoretically neutral, but presuppose a frame of reference that can and should be made explicit. Those who are concerned with building models, meanwhile, must acknowledge the importance of experimental testing and of reinterpreting experimental data based on different points of view. The key to progress in understanding prosody is to examine assumptions and, as far as is possible, test them empirically. That is the single most important point that we hope will emerge – both by example and by juxtaposition of examples – from the contributions collected here.

1.4 The Contributions

The foregoing sketch allows us to place the individual contributions to this volume in a common context. The first three – by Gårding, Thorsen, and Ladd – deal with the representation of intonation, and they nicely illustrate the range of approaches we have discussed so far. Thorsen accepts many of the assumptions of the "concrete" approach, Ladd's model is rather "abstract", and Gårding falls somewhere in between, though perhaps closer to Thorsen in most respects. Gårding presents a summary of the Lund model of prosody, and illustrates how it can be used to generate different prosodic characteristics in three different languages. Pitch is specified by the interaction of grid lines for sentence intonation, tone sequences for accent, and various modifications for rhythmic and other effects. The grid lines typify the intermediate position that Gårding takes: their height and shape directly represent the pragmatic effects of intonation, but they also play a role in specifying the phonetic detail of the rather abstract tone sequences by which Gårding represents accent. Thorsen's model is more "concrete", representing both sentence intonation and accent by interacting configurations. The main concern of her paper is to demonstrate the necessity for such a model in Danish; in particular, she argues that the more abstract notions of sentence stress and tone sequence are unable to account for her instrumental data. Ladd, finally, illustrates a feature analysis of tone sequences, and argues that such a relatively abstract description can actually give more insightful and more accurate accounts of overall slope than models in which overall slopes are directly specified as lines or configurations.

The next three contributions, by Vaissière, Brown, and Cutler, are all concerned in one way or another with prosodic function. Vaissière explicitly adopts the "concrete" point of view and the assumption of a direct link between function and form, arguing that prosodic function is to a great extent physiologically natural. She discusses several specific functions – marking stress groups, marking phrase boundaries, marking questions, and so on – that appear to be found in a wide variety of languages. For each such function she identifies common acoustic correlates and emphasizes the importance of physiological data. Brown, by contrast, assumes a relatively abstract approach to prosodic function and concentrates on a specific problem, namely the role of prosody in signalling discourse structure. She treats a single formal category, pitch prominence, and identifies it with a functional label, "new information". In keeping with the assumption of indirect connection between prosodic form and utterance interpretation, she is at pains to show that there are *not* intonational correlates for the whole range of possible discourse "statuses" posited by E. Prince [1981] and others: pitch prominence is a "generalized 'pay attention' marker", and more specific categorization of discourse status is inferred by the speaker on the basis of a variety of syntactic and other cues. Cutler, finally, reports on her own corpus of speech errors involving prosodic features. Different types of errors produce different patterns of correction; in general, speakers are more likely to correct prosodic errors the more seriously they disrupt the interpretation of the utterance. Other details suggest that language users behave as if prosody were definitely a peripheral, separate, and in many cases dispensable component of language.

The final three contributions are all relevant to recent abstract models of prosodic structure proposed by Selkirk [e.g. 1980] and others. Hirst in particular is concerned to show that even research on fine phonetic detail may be relevant to the broadest theoretical questions of the organization of language. He shows how, within the general framework of Selkirk's model of prosody, it is possible to describe intonation in a way that represents both its relation to the hierarchical organization of utterances and its phonetic linearity. His linear phonetic representation bears a considerable resemblance to

Gårding's, while his hierarchical structures are essentially the same as those discussed at greater length by Nespor and Vogel.

Buxton describes several experiments in which fine details of speech timing were manipulated in stimuli. Her results point clearly to the existence of temporal structure that is critical for the fluent processing of speech. Among other things, these results can be read as evidence for the centrality of prosodic structure in language: timing is not simply a matter of, for example, adding length to a syllable because of an accompanying feature of stress, but rather of fitting all the sounds, all the syllables, into an overall frame that is part of the phonological representation of the utterance.

This view is given a more explicit form by Nespor and Vogel, who build on Selkirk's model of prosody and make predictions about prosodic disambiguation in Italian. They also report experimental results that bear out their predictions. What is of particular relevance to the question of the centrality of prosody is the fact that the prosodic structures which Nespor and Vogel discuss affect many different physical aspects of the speech signal, including phenomena such as "radoppiamento sintattico" which are not normally considered "prosodic". That is, their model implies that it is not possible to identify certain functions or acoustic features as prosodic and treat them apart from the rest of language, but rather that everything in the realization of an utterance is potentially affected by its prosodic organization.

1.5 The Purpose of an Interdisciplinary Volume

While we have argued strongly that different approaches to studying prosody reflect empirical assumptions that can be stated and examined, we do not mean to underestimate the extent to which different traditions of inquiry are due simply to lack of communication among linguistics, phonetics, and psychology. Traditional disciplinary boundaries still make themselves felt most obviously in the fact that, for example, people who read Linguistic Inquiry do not normally read The Journal of Experimental Psychology, and vice versa. The effects of this can be illustrated most dramatically in the very recent history of work on rhythm, timing, and metrical organization. Within linguistics, an entire school of prosody and metrics has developed within the last five or six years, based on Liberman and Prince's article "On Stress and Linguistic Rhythm" [1977]. (Notable works in this tradition include Selkirk [1980], Kiparsky [1979], Hayes [1981], and McCarthy [1979].) Meanwhile, in psychology and experimental phonetics, in about the same length of time, there has been a resurgence of interest in the question of "isochrony" and rhythm, as represented by the work of Martin and his colleagues [e.g. Martin, 1972; Meltzer et al., 1976], the work on perceptual centres by Morton et al. [1976], and other work on timing by Lehiste [1977], Fowler [1979], Donovan and Darwin [1979], Nakatani et al. [1981], and Scott [1982]. Despite the fact that both these approaches are directly concerned with the temporal patterning of prominence in language, however, there has been virtually no contact between them. Linguistic model building has concentrated on the assignment of lexical stress, and has worked with notions like "relative prominence" and "metrical grid" without making any attempt to show what those might mean in perceptual or acoustic terms. The psychological work, meanwhile, strongly suggests the existence of temporal structure to speech, but has scarcely gone beyond formulations like "a tendency to isochrony" in explaining experimental results. (Martin [1972], it is true, proposes a hierarchical model not unlike Liberman and Prince's, but his experiments have not really attempted to test its predictions.)

One obvious purpose of a volume like this, then, is to acquaint people who read one set of journals with work of the sort that appears in various other sets. Yet even when researchers from one field explore what has been done in another, the problems of methodology, research strategy, and background assumption remain. The thousand subtle lessons that are learned in becoming a practitioner of one field may contribute to the feeling that work in neighboring fields is uninteresting, irrelevant, or worse. The tension between model building and doing experiments – which in this context strongly tends to divide the linguists from the psychologists and the phoneticians – is still often a significant barrier to the spread of ideas.

Beyond the obvious goal of making papers from different disciplines available in the same place, our main aim in assembling this collection has been, as we said, to illustrate the interdependence of models and measurements. Specifically, we hope to have raised questions like the following:

a) Vaissière presents data showing the universality of certain prosodic features, while Gårding and Thorsen both present data showing how language-specific prosody can be. What sort of model can account for both kinds of observations?

b) We have described both Ladd and Brown as exemplifying the "abstract" approach to prosodic function. Why, then, are the details of their models so different, and what predictions do their models make that could be tested empirically?

c) Buxton suggests certain ways in which the hierarchical model assumed by Nespor and Vogel and Hirst may be inadequate to account for data from speech perception. In what ways could the hierarchical model be revised to take account of such data? How would revision affect the model's applicability to broader questions of the organization of language? How could a revised model be tested?

d) Cutler discusses the problem of uncorrected prosodic errors, while Brown's emphasis on "speaker choice" would appear to make it difficult to identify a prosodic error unless it is corrected. If we grant the difficulty of directly testing the validity of the concept "uncorrected prosodic error", then we must instead test larger models of which that concept is or is not a part. What would such models look like, and how would we test them?

If readers are asking themselves such questions when they finish this book, then it will have fulfilled the best purpose for projects that cross traditional boundaries between fields – to enable people on opposite sides of such boundaries not just to become aware of each other's work, but to see the point of it as well.

2. A Generative Model of Intonation

Eva Gårding

With 8 Figures

In this contribution I will do four things:
1. Give a brief presentation of an intonation model developed at Lund[1]
2. Apply it to a sentence in Swedish, Greek and French
3. Use it to compare the three prosodic systems
4. Indicate how it can accommodate pragmatic features in prosody

2.1 The Model

Let me first state the most important assumptions behind the model. It is based on an analysis which separates lexical prosody from phrase and sentence prosody. The input to the model is a sentence, equipped with markings for lexical accents or tones, accents (tones) at phrase and sentence level, morphological and phrase boundaries, and the mode of sentence intonation. We assume that all these factors combine and interact to produce the final temporal and tonal pattern of the actual speech signal[2]. The model simulates this process in a number of steps. These steps are not intended to simulate the process of the production of actual speech. The model should be seen as an attempt to systematize the prosodic part of language description. It may have applications for studying language learning [Gårding, 1981] and language acquisition.

Figure 2.1 shows some examples of fundamental frequency curves from a sentence in Swedish, Greek and French. (The English translation is "Madame Marianne Mallarmé has a mandolin from Madrid".) The three sentences are comparable in pragmatic, semantic, syntactic and phonetic structure. The corresponding inputs to the model are given in the lower part of the figure.

---

[1] The original model was developed in collaboration with Lindblad [Gårding and Lindblad, 1973]. Important contributions were made by Bruce [1977] and by Bruce and Gårding [1978]. For recent work with the model see Gårding [1981] and Bruce [1982]. The analysis of Greek and French presented here was done in collaboration with Botinis and Touati [Gårding et al., 1982].

[2] We also assume that intonation can be successfully generated with fundamental frequency (pitch) only. We intend to analyze cases where fundamental frequency and intensity differ.

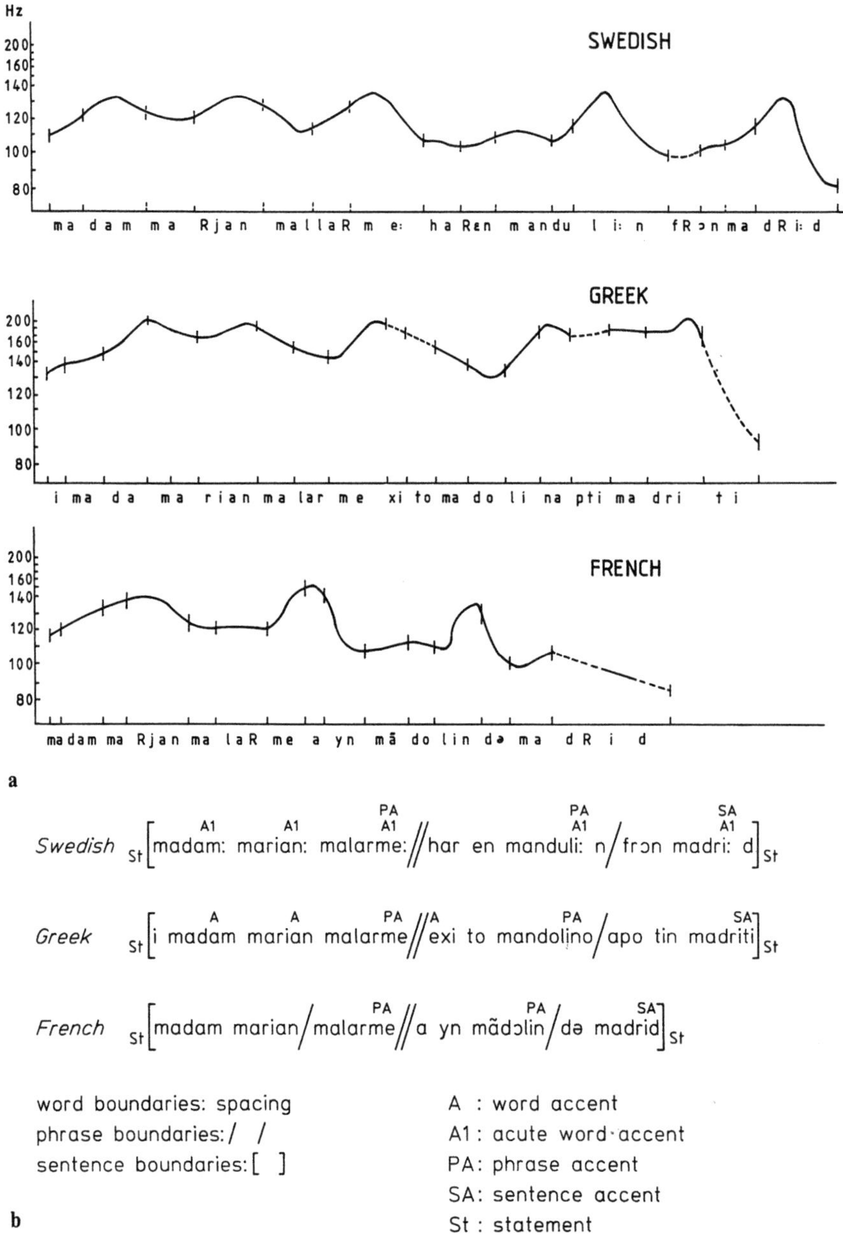

Fig. 2.1. (**a**) Statement intonation in observed sentences; (**b**) inputs to the model [Gårding, 1981]

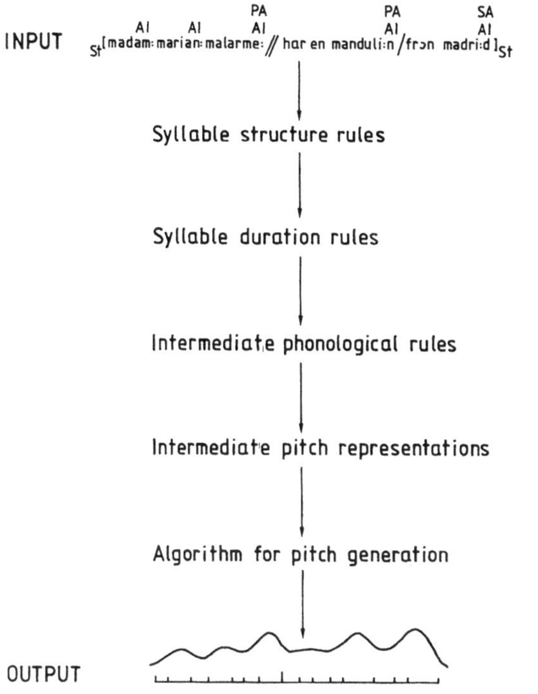

Fig. 2.2. Model for prosody [Gårding, 1981]

Figure 2.2 gives the stages of the model. For each stage there is a set of conventions and prescriptions. Conventions are general rules common to all languages, and prescriptions are language-specific or dialect-specific rules. This arrangement is convenient for cross-linguistic comparisons.

In this presentation of the model I will concentrate on the intonation part and assume that the sentence has passed through the first two stages, where it has been given the correct syllable durations [Gårding, 1981]. The intermediate phonological rules which represent the next stage are needed to account for cases where the input marks have an effect on duration only and not on pitch. One such rule which is fairly general across languages is that accent marks after the SA (sentence accent) expressing focus are deleted[3]. Yet the accent contrast is still there. It may now be carried by the durational feature which was generated by the previous rule. Another rule common to the languages we have studied is that in a sequence of accents within a phrase, the middle ones are deleted. (For the consequences of this rule see below.)

The rules of the fourth stage, the intermediate pitch representations, convert the abstract input marks into more concrete symbols. Global features, which pertain to sentence and phrase intonation, are expressed as a rise, a fall or a level, or as a combination of these elements. Local features,

[3] This rule applies only partly in Swedish [Bruce, 1977; Gårding, 1981].

Table 2.1. Algorithm for generating pitch contours

Rule 1.	*Sentence and phrase intonation* Draw the tonal grid using sentence type and major phrase boundaries
Rule 2.	*Sentence and phrase boundaries* Insert highs and lows on the grid according to language and dialect
Rule 3.	*Sentence and phrase accent* Insert highs and lows on the grid according to language and dialect
Rule 4.	*Word accent* Insert highs and lows on the grid according to language and dialect
Rule 5.	*Contrastive word accent* Adjust highs and lows according to language and dialect
Rule 6.	*Context rules* Adjust highs and lows according to context
Rule 7.	*Concatenation* Connect neighboring generated highs and lows

which pertain to syllables or words, are expressed as highs and lows or as combinations of highs and lows, which become turning points in the final pitch contour. The specifications of highs and lows may differ even within the dialects of a language. This difference, combined with the accent deletion rule above, may give rise to different phrasal intonations – a "hat pattern" [Cohen and 't Hart, 1967] in one language or dialect as compared to a "trough pattern" in another.

The last stage of the model is the pitch algorithm. It consists, in its most recent version, of seven rules (Table 2.1). Rule 1 gives instructions for the tonal grid, the global frame for the sentence intonation within which the local pitch movements can develop. Rules 2, 3 and 4 insert highs and lows on this grid. Rule 5 concerns contrastive accent. A very common result of this rule is that the highs and lows are adjusted on the frequency scale (and sometimes also on the time scale) to make room for the larger pitch movements associated with contrast. Rule 6 takes care of influence from the prosodic and segmental context. Rule 7, finally, concatenates all the generated highs and lows with a smooth line over the voiced segments.

Some of these rules need special comment, for instance the grid. Figure 2.3 shows an idealized grid for a declarative sentence that has the same structure as the sample input sentence. The vertical scale is logarithmic and covers one octave. For synthetic speech the grid must be specified numerically. The interior solid lines represent the bounds of normal, not weakened, accents within a phrase. The exterior broken lines represent the floor and the ceiling of the voice in the normal register. These exterior lines are used for semantic and pragmatic effects, for strengthened accents, for instance in combination with focus. We are less sure about the course of these lines, particularly the upper one.

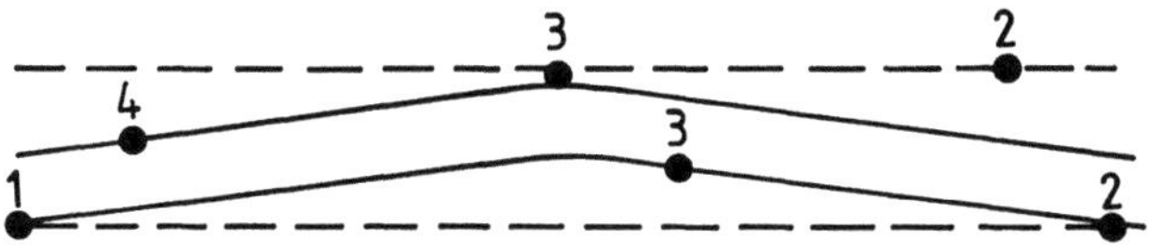

Fig. 2.3. Example of grid with insertion of highs and lows [Gårding et al., 1982]. (**1**) Grid: sentence and phrase intonation, (**2**) sentence accent: high and low, (**3**) phrase accent: high and low, (**4**) word accent: high

In the grid the pivot marks the boundary between, in this case, the subject phrase and the predicate phrase. Switching the order of the constituents to put the predicate as the first phrase and the subject as the last shows that the rising contour is tied to the first phrase and the fall to the last, not to a particular constituent.

So far the grid has been worked out for one particular sample of sentences with focus placed in one of three positions (*Mallarmé, mandolin* and *Madrid*). We know from earlier work that the slope of the fall of the interior lines is to some extent dependent on the length of the phrase. According to Bruce [1982] the length of the phrase also has an effect on the frequency value of the pivot. Similar results have been obtained by Thorsen for Danish [1980 c] and by Touati for French [unpublished].

The context rules, 6 A and 6 B (Table 2.2), are also needed to generate the observed pitch pattern of Fig. 2.1. *The Preparatory Rule,* 6 A, prepares for a subsequent high or low by inserting a point of the opposite value. In a given surface contour, then, a criterion is needed to determine which turning points are preparatory and which are not. For instance, in a pitch curve moving from high to low to high, the low can be considered either as a preparatory low for the following high or as a low prepared for by the preceding high. In the diagram below, parentheses indicate points generated by the preparatory rule.

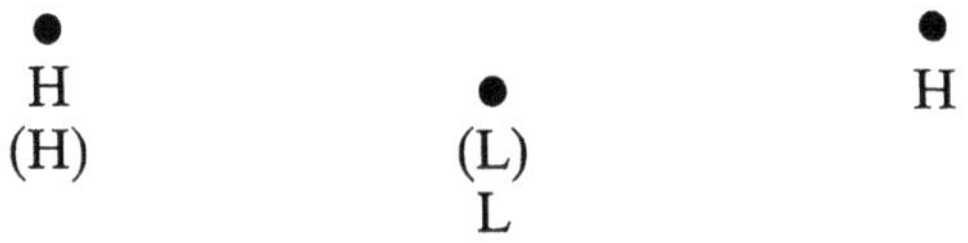

The choice between such alternatives is made by studying a pitch contour in varying context and tempo. It turns out that some points are fairly constant in relation to a certain segment, while others vary or disappear (Fig. 2.4).

Table 2.2. Rules of pitch algorithm generating declarative sentences consisting of two major phrases in narrative style

Rule		Swedish (South)	Greek (Athens)	French (Standard)
1 SI PI	Sentence and phrase intonation	Rising-falling grid	do	do
2 SB PB	Sentence and phrase boundary	SB: Low . . . low PB: Low . . . low	SB: Low . . . low PB: Low . . . high	SB: Low . . . PB: Low . . . WB: Low . . .
3 SA PA	Sentence and phrase accent	SA: High-low PA: High-low	SA: High-low PA: High	SA: Low PA: High
4 WA	Word accent	High	High	Void
5 CA	Contrastive accent	High-low with large range	High-low with large range	Low-high for first syllable of word
6	Context modifications A. Preparatory rule B. Assimilation rule etc.	A. Preparatory rule B. Assimilation rule	do do	do do
7	Concatenation	Connect neighboring points by a smooth line over the voiced segments	do	do

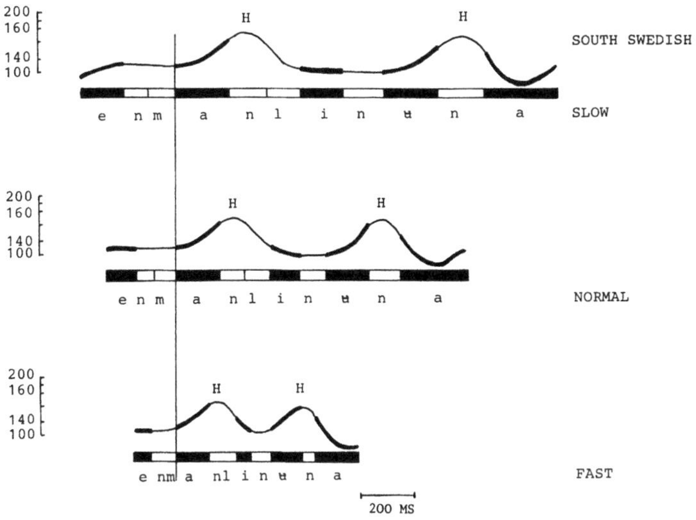

Fig. 2.4. Effects of tempo differences on turning points [Gårding, 1975b]. Average durations and fundamental frequency curves are shown for the phrase *en manli nunna* (marked A2 A2) "a masculine nun", spoken at three different rates. The high turning points of the accents are fixed relative to the segments, but in fast speech the preceding low turning points come closer to the beginning of the phrase. This motivates the choice of high as the pitch representation of the South Swedish accents

The constant points, or "fixed points", as they were called by Gårding [1975 b], are chosen as the pitch representation.

The Assimilation Rule, 6 B, expresses undershoot phenomena. Before a steep SA fall, the preceding low is undershot in Swedish and Greek. One possible reason is that the steep fall is a strong manifestation of the accent and therefore does not need a full preceding preparation.

Priority Rules are context rules that indicate priorities when the results of two rules clash. For instance, when in Swedish or Greek a phrase or sentence accent is initial, there is a conflict between the low required by the phrase boundary and the high needed for the accent. In this case the accent high is given priority.

The Concatenation Rule, 7, is a mechanical rule which is the same for all three languages. It accounts for "hats" or "troughs" formed between accented syllables, and characterizes the way in which a group of unaccented syllables forms a prosodic unit with the preceding or following accented one.

2.2 Application to Swedish, Greek and French

Table 2.2 contains the information necessary to apply the rules of the pitch algorithm to Swedish, Greek and French declarative sentences. All the rules except Rule 1 remain unchanged in other contours. The table has some redundancies. Rules 2 and 3 overlap when a phrase or sentence accent is final. In the case demonstrated here there is no conflict. The rules give the same result. In other cases, however, priority rules may be needed.

Figure 2.5 illustrates how the points given by the various rules are inserted into the grid which has been generated by Rule 1. The numbers 2, 3, etc. refer to the corresponding rules of the algorithm. Note that the lengths of the grids have been adjusted to the observed patterns. The match between the observed curves of Fig. 2.1 and the curves obtained by interpolation in Fig. 2.5 is satisfactory. (For a fuller account see Gårding et al. [1982].)

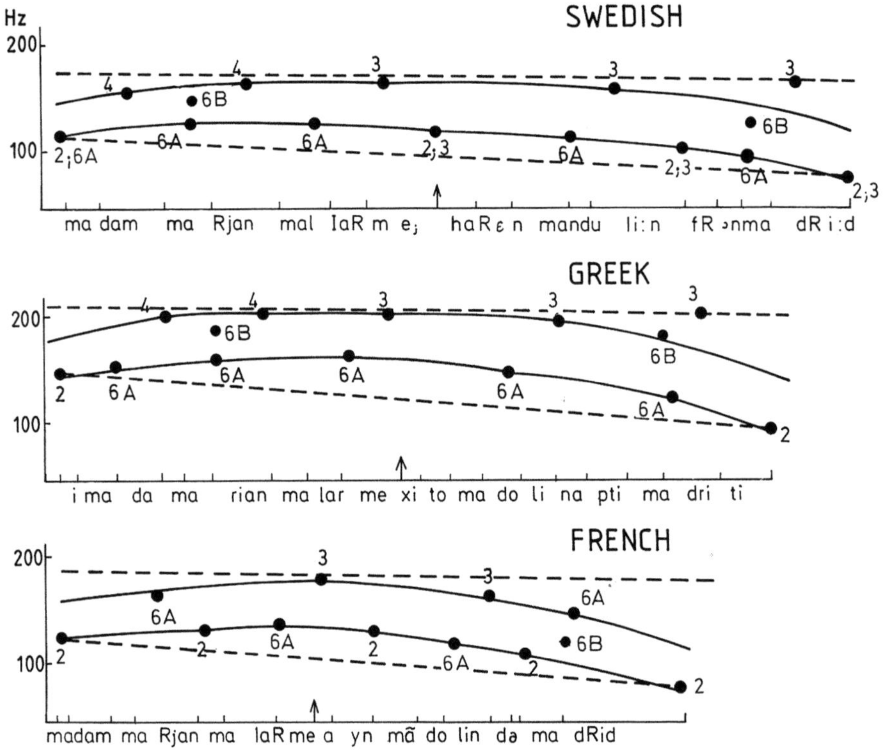

Fig. 2.5. Generated highs and lows inserted on tonal statement grids for Swedish, Greek and French [Gårding et al., 1982]

2.3 Comparison

2.3.1 Intonation and Boundaries

Apart from differences in overall length, the grids are similar, with a rise of the interior lines over the first phrase (here subject) and a fall over the second (here predicate). The pivot is connected with the boundary between these two constituents, but in different ways. In the dialect chosen to represent Swedish (Skåne), this boundary is low; in Greek and French it is high.

For the other boundaries, the French speaker uses low to mark the beginning of a phrase. To mark the end of an internal phrase, he uses high on one of the upper lines, with the highest line for the largest constituent. This is in agreement with the analyses of others, e.g. Martin [1975] and Vaissière (this volume).

2.3.2 Accentuation

Since all accents in French are phrase or sentence final, it would also be possible to analyze them as boundaries. This has been done by Vaissière (this volume).

Instead of the phrase-internal accents characteristic of Swedish and Greek, French has phrase-internal boundaries, marking the beginning of semantically important words with low.

For the Swedish and Greek dialects that we have investigated, high seems to be the obvious choice for the representation of the word accent. In the present comparison with French and Greek, only one of the word accents (A 1) has been used for Swedish. For A 2 the same representation will be used, with the added specification that it has a later timing, corresponding to about half an accented syllable. With this analysis, the representation reflects a constant pitch feature in connection with the accented syllable[4].

Table 2.3 shows the pitch representations proposed for the linguistic variables of the three languages. It gives an important part of their prosodic structures and sums up the main conclusion of our comparison: phrase and sentence intonation are similar in declarative sentences, while accentuation and boundaries differ.

---

[4] It can be seen in Fig. 2.1, for example, that the low of the Swedish accent manifestation is not tied to a specific segment. It is therefore generated by the preparatory rule or the phrase accent rule. This does not contradict our earlier analysis of Swedish dialects [Bruce and Gårding, 1978], which highlighted the common structure of the accent manifestations of the dialects. High-low was chosen as the common denominator for all Swedish accents, with the timing of the high-low as the distinctive feature of the two accents on the one hand, and as a dialect marker on the other.

Table 2.3. Summary of pitch representations for accents and boundaries in declarative sentences. The letters in parentheses are consequences of the preparatory rule; the vertical arrows indicate widened range; and the horizontal arrow indicates delayed timing

	SB	PB	SA	PA	WA	CA
Swedish (South)	L ... L	L ... L	↑ HL ↓	HL	1. H(L) ⟶ 2. H(L)	↑ HL ↓
Greek	L ... L	L ... H	↑ HL ↓	(L)H	(L)H	↑ (L)H ↓
French	L ...	L ...	(H)L	(L)H	–	(L)H

This analysis happens to be consistent with some impressionistic statements about the languages. The South Swedish dialect is described by other Swedes with rising accents as slow, monotonous, chopped up and down to earth. The falling phrase accents, together with the fact that Skåne is a recently conquered province (1658 AD) may be responsible for these impressions. Greek uses rising pitch ((L)H) as a linguistic variable much more than Swedish. This, combined with its preference for open syllables, may explain why it has a hammering, insistent effect on a Swede.

The most conspicuous pitch characteristic of French to a Swede is the recurring phrase-final rising pitch, and this is what Swedes first imitate, regardless of their dialectal background. This and other characteristics of French spoken by Swedes will be treated in a dissertation by Touati. When Swedes speak English, on the other hand, they tend to use Accent 1 for all accented syllables with the manifestation typical of their dialect [Gårding, 1962].

The real importance of an analysis like that in Table 2.2 is that it summarizes the pitch representations and therefore the means of tonal expression of a language. By adding an account of the function and distribution of the prosodic features needed in the input and the intermediate phonological rules, we obtain the main part of a prosodic grammar. The uniform way in which such grammars can be obtained makes prosodic comparisons between languages meaningful, and thus they will also be efficient aids in analyzing foreign accents and teaching pronunciation [Gårding, 1981].

2.4 The Model Applied to Questions and Pragmatic Effects

Figure 2.6 shows the application of the pitch rules to a statement and two kinds of questions [Gårding, 1979]. Question 1 is a yes/no question with inverted word order, and Question 2 is a yes/no question without inversion. The figure shows that it is the grid that changes when we go from a statement to Questions 1 and 2. In particular, the change affects the scale for the auxiliary lines. However, this scale is not changed uniformly. The upper topline, the ceiling, is raised more than the other lines, while the baseline, the floor, remains in its original position[5]. With an expansion of the total register the interior lines are straightened out and compressed. The difference between the intonation types is thus a difference in the grid, whereas the rules that generate the highs and lows pertaining to the accents remain the same, except for an optional rule producing a terminal rise in questions. The timing of the accentual highs and lows seems likewise to be a very constant feature.

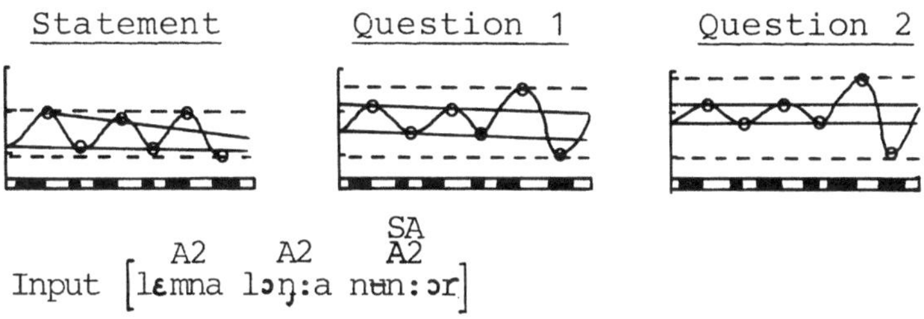

Fig. 2.6. Result of applying the pitch algorithm to a statement and two kinds of questions. (Note: Simple verb phrase only without pivot) [Gårding, 1979]

Data from a recent study [Gårding, 1982] show that accentual highs and lows remain fixed even when the intonational register is compressed or expanded in varying pragmatic contexts. Figure 2.7 shows different intonations of the sentence *Hon gick inte o(ch) la sej* "She didn't go to bed", elicited in six different contexts[6]. These contexts are defined by the content of the introductory sentence given in Table 2.4. Depending on the context, the intonation of Sentence 2 is classified as neutral, downgraded or upgraded relative to Sentence 1. The column to the right gives a classification of Sentence 2 in text-linguistic terms. In the comments that follow I shall let sentence 1.2 denote Sentence 2 of Text 1 and so on.

---

[5] This observation is consistent with the model presented by Pierrehumbert [1980], which also incorporates the concept of an invariant baseline, although it should be noted that Pierrehumbert's baseline concept differs from the one described here.

[6] The aim of the experiment was to see if listeners could match sentences with their contexts. In most cases the task was easy.

Table 2.4. Intonation in texts consisting of two sentences

	Text unit		Manifested class of intonation		Relation of intonations
Text	Sentence 1	Sentence 2	Sentence 1	Sentence 2	Sentence 1/Sentence 2[a]
1	*Hon var inte trött* "She wasn't tired"	*Hon gick inte o(ch) la sej* "She didn't go to bed"	Neutral	Neutral	Additive
2	*Hon var trött* "She was tired"	*Hon gick inte o(ch) la sej*	Neutral	Upgraded	Adversative
3	*Hon SA* "She SAID"	*Hon gick inte o(ch) la sej*	Upgraded	Downgraded	Complement
4	*Hon var alldeles säker på sin sak* "She was quite sure"	*Hon gick inte o(ch) la sej*	Neutral	Upgraded	Assertive
5	*Hon gick ut* "She went out"	*Hon gick inte o(ch) la sej*	Neutral	Upgraded	Contrastive
6	*Hon hade hög feber?* "She had a high temperature?"	*Hon gick inte o(ch) la sej?*	Neutral	Neutral	Additive

[a] The terms in this column denote the second element as compared with the first.

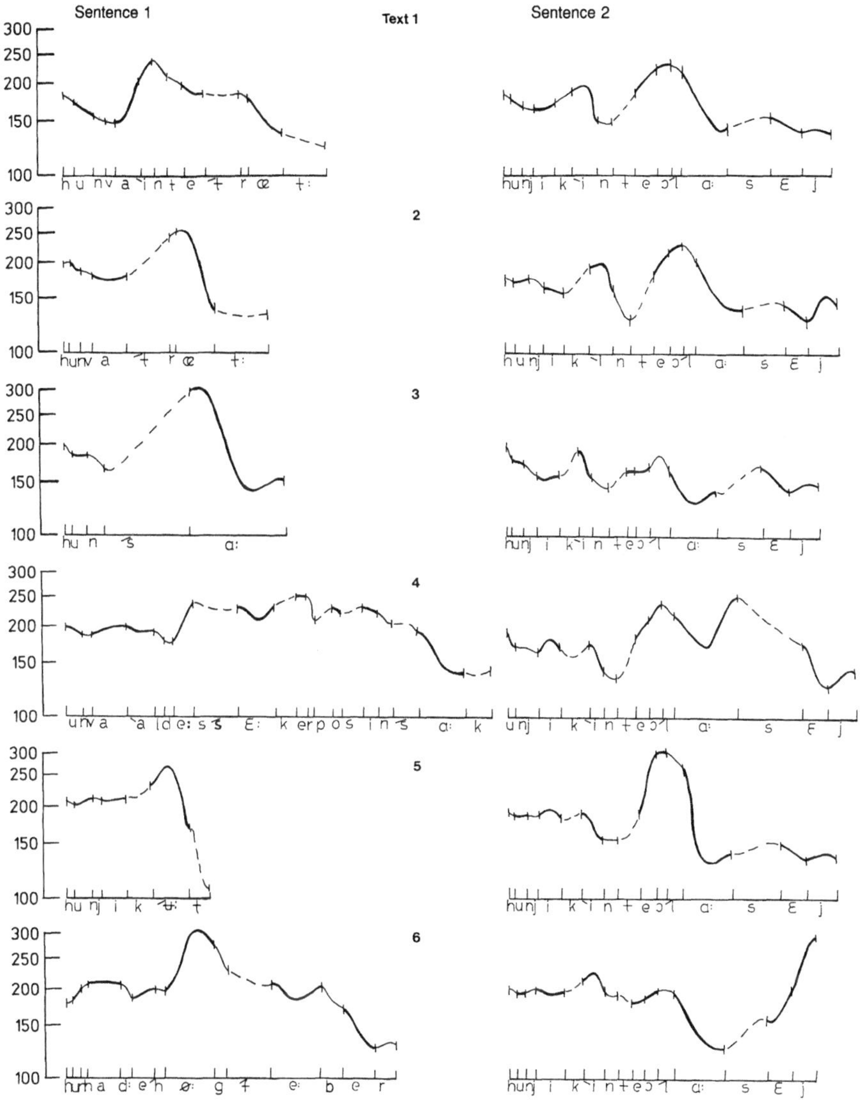

Fig. 2.7. The pragmatic function of intonation illustrated by pitch contours derived from texts consisting of two sentences

The following observations can be made on the basis of these data. As before, the accentual points are fixed relative to the segments in all cases, even with very strong compression and expansion, as in 3.2 and 5.2.

It is also possible to analyze the contours in terms of grids. Tentative grids consisting only of the interior lines are shown in Fig. 2.8. One falling grid seems to be sufficient to generate the intonation of Sentences 1.2, 2.2, 3.2

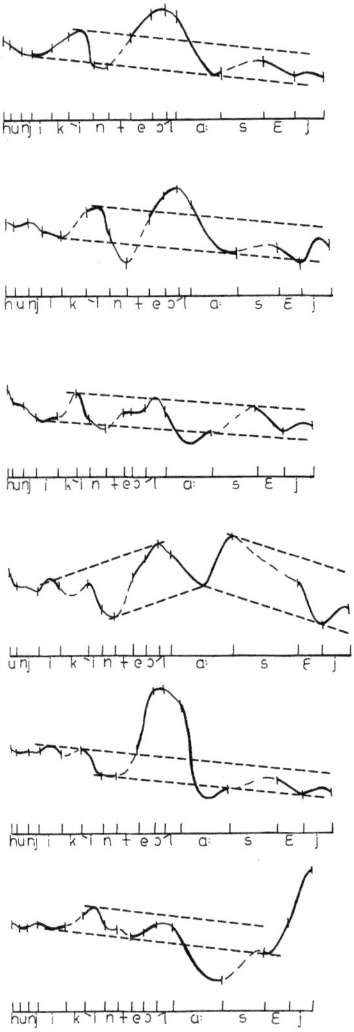

Fig. 2.8. Tentative grids for contours of Sentence 2, Fig. 2.7

and 5.2. Sentence 1.2, produced as an added comment to 1.1, reaches the lower baseline only at the end. In Sentence 2.2, with an intonation adversative to 2.1, the lower baseline is reached at *in,* thereby creating a larger range for the pitch movements in connection with the negation *inte.* Sentence 3.2 is subordinate to the focused predicate of 3.1. The speaker seems to achieve this effect by avoiding the upper topline and hence remaining in the shadow of the sentence accent of 3.1. In 5.2 the pitch movements in connection with the dominant accent (SA) on *inte* and the following fall on *la* dominate the contour, and the rest is compressed. As in the questions, the distance between the upper toplines is expanded both downwards and upwards.

The intonation of 6.2 may fit into a falling grid with a local rise to the uppermost topline from the baseline low of the word with SA. This intonation is similar to the "valley tune" of English [O'Connor and Arnold, 1961] typical of yes/no questions. At first glance a Swedish contour looks more complicated, owing to the local ups and downs created by the word accents.

Sentence 4.2, finally, is interpreted as having a rising-falling grid similar to that of Fig. 2.3. There are two strong accents, on *inte* and *la,* and the direction change of the grid occurs on the latter. The unaccented onset *hon gick* "she went" is on a fall. This contour was intended to express assertion. It seems to be a repetition of the preceding contour, 4.1, with an expanded range over the negation and the predicate. This expansion is created by means of the accentual pitch movements.

The preceding analysis shows that the model can account for various pragmatic effects. The intonation of a falling grid, for instance, may be varied by letting the accents of a word or a phrase reach or not reach the outer lines of the grid. One may speak of compression when a curve stays between the interior lines and of expansion when the grid is widened and the exterior lines are used for prominent accents and for boundaries.

The use of expansion and compression is probably an almost universal feature of intonation. In a language like Swedish or Greek these effects are achieved mainly by changing the ranges of the accents, probably in a predictable way. In a language without accents (like French, according to some analyses) this function is fulfilled primarily by a modification of the boundaries.

3. Two Issues in the Prosody of Standard Danish

Nina Thorsen

With 7 Figures

3.1 Introduction

As indicated by the title, this contribution falls into two parts. The first deals with the lack of an obligatory sentence accent in Standard Danish, a phenomenon which, to judge from descriptions of more or less closely related languages, is somewhat exceptional. The second concerns the tonal sequence representation suggested by Pierrehumbert [1980] and by Ladd in his contribution to this volume.

3.2 Sentence Accent

Sentence accent, primary accent, sentence stress, nuclear stress, tonic, focal accent, Satzakzent, etc., are more or less synonymous terms which designate the one stressed syllable which is more prominent than other stressed syllables in a stretch of speech.

The works which implicitly or explicitly assume the existence of a nuclear stress in British and American English are too numerous to permit an exhaustive listing. In a review of past work on prosodic features, Crystal says [1969, p. 207]:

> There is general agreement about the internal structure of the tone-unit in English. Minimally, a tone-unit must consist of a syllable, and this syllable must carry a glide of a particular kind. This is the obligatory element, and is usually referred to (in the British tradition) as the *nucleus* of the tone-unit. ... The presence of a nucleus is what accounts for our intuition of "completeness" at the end of the unit; if it is omitted, the auditory effect is one of "being cut short".

Liberman and Prince [1977, p. 259] – in a theoretical framework which differs from most previous descriptions of English stress – also assume a main stress which is the most prominent terminal element of a given constituent and is termed a "designated terminal element".

Halliday [1967 b] rather categorically associates the tonic with "contrastive" or "new" in the information structure. Experiments reported in Currie

[1980, 1981] and Brown et al. [1980] demonstrate that this claim is far too strong: listeners are not unanimous in their tonic assignment (neither in read nor in spontaneous speech) and (consequently) a perceived tonic does not necessarily fall on the "new" information in the utterance. Brown et al. [1980, p. 157] suggest that

> in sentences read aloud there will generally be some extension of pitch movement on the last stressed syllable. This cannot be taken as marking new information, since all items in the sentence must be new. What it does mark, we suggest, is "sentence final",

which ties in well with the quotation from Crystal [1969] above.

Bolinger's [1958, 1972] account of (American) English pitch accents does not include the concept of a sentence accent, and at first glance his analysis resembles that of Dutch (cf. below). However, Bolinger [1972] is perhaps rather more categorical in the association between information structure and pitch accents than 't Hart and Collier [1979] and Terken [1980 a, b]. Furthermore, Bolinger [1958] clearly implies the possibility of more and less prominent pitch accents by referring to "the stress ... that is the most prominent one in the utterances tested" [p. 113] and "the major stress" [p. 116]. At this point, though, I wish to point out that there is a certain difficulty in comparing descriptions across languages and authors, because the pragmatic constraints (if any) of the materials analyzed are not always made explicit, and in any case are rarely identical across investigations.

The existence of a Satzakzent in German which falls on the most important word in the utterance seems to be uncontested [von Essen, 1956; Stock, 1980, pp. 79–80; Klein, 1980]. Sentence accent is also an undisputed reality in Standard Swedish [Bruce, 1977; Gårding, 1980]. Bruce is not categorical in the association of sentence accent with "new" information, but states,

> It seems to me that the Functional Sentence Perspective – the theme/rheme distinction – can account for a great deal of what has conventionally been attributed to emphasis and contrast

[Bruce, 1977, p. 21]. According to Carlson et al. [1975, p. 212], prosodically neutral utterances, i.e. utterances with a "non-contrastive (focus-free) stress pattern" are possible. (Note that they seemingly equate contrast and focus.)

The relation between sentence accent and information structure may not be straightforward, and there may also be a marked difference between planned and spontaneous speech with respect to the number and placement of sentence accents. I do not think that it stretches the descriptions of Standard Swedish, English and German too far, however, to conclude that in pragmatically neutral speech the last accented syllable in the phrase (or whichever the relevant chunk of speech may be) will normally be more prominent than preceding stressed syllables, a prominence which is realized by a greater duration of and a more elaborate fundamental frequency (F_0) movement within or in the immediate environment of that syllable. Under different circumstances the context may evoke an earlier location of this prominence, for instance under conditions of contrast.

In Standard Danish, in pragmatically and emotionally neutral speech (as, for instance, in the reading of context-free utterances), *none* of the stressed syllables is more prominent than the others. Such prominence is not evident in acoustic registrations (in duration and/or F_0 contour), nor is it present auditorily. There is nothing incomplete about such prosodically neutral utterances, and although they may not be very frequent in spontaneous speech, they certainly do occur, they are not unnatural, and they are very easy to elicit from speakers in a reading situation. I should add that this is a completely uncontroversial fact among Danish scholars. It appears, then, that three degrees will suffice to account adequately for the distribution of stresses in Standard Danish: main stress (assigned to the lexically stressed syllables in most non-function words), secondary stress (assigned to the second – and following – lexically stressed syllables in compounds), and weak stress [Basbøll, 1978; Heger, 1981, pp. 118–119, 122; Rischel, 1972, 1975, 1980, in press]. In similar utterances in Standard Swedish, English, and German a fourth degree is required: the sentence accent.

In Thorsen [1980 a] I report the results of acoustic analyses of emphasis for contrast in Standard Danish, which may be summarized briefly as follows: When emphasis for contrast is introduced in a sentence, the stressed syllable of the emphasized word will stand out clearly from the surroundings. This is brought about by a rising of F_0 (except in initial position), an elaborate F_0 rise within that syllable, and a deletion of the F_0 deflections (the rises from the stressed syllables) in neighbouring stress groups, to the effect that the immediate surroundings, except the first post-tonic, fall away sharply from the stressed syllable of the emphasized word (Fig. 3.1). (Statements and syntactically unmarked questions differ with respect to the extent of the influence of emphasis on the prosodic patterns: F_0 patterns two stress groups away from the emphasized word retain – at least partly – their rises to the post-tonic in statements but not in questions. It may be that more of a change is invoked in the prosodic patterns on marked intonation contours – which accompany syntactically unmarked questions – than on unmarked contours, i.e. terminal declarative sentences). For the sake of clarity I will disregard the limited influence of emphasis in terminal declarative sentences and conclude: Short utterances with emphasis for contrast reduce *tonally* to one prosodic stress group in the sense that only one *low* + *high-falling* pattern occurs in them. The difference between statements and questions with contrast emphasis is mainly located in the movement within the emphatic syllable which rises higher in questions, and in the post-emphatic syllables which run higher in questions than in statements.

Clearly, then, Danish contrast emphasis shares some features with the acoustic manifestation of sentence accent in other languages. It is difficult to carry this comparison much further on a general basis, because there are differences between English, German and Standard Swedish: Satzakzent in German is not signalled by any special F_0 movement within the stressed syllable (except if it is the last one in the utterance), but rather by what

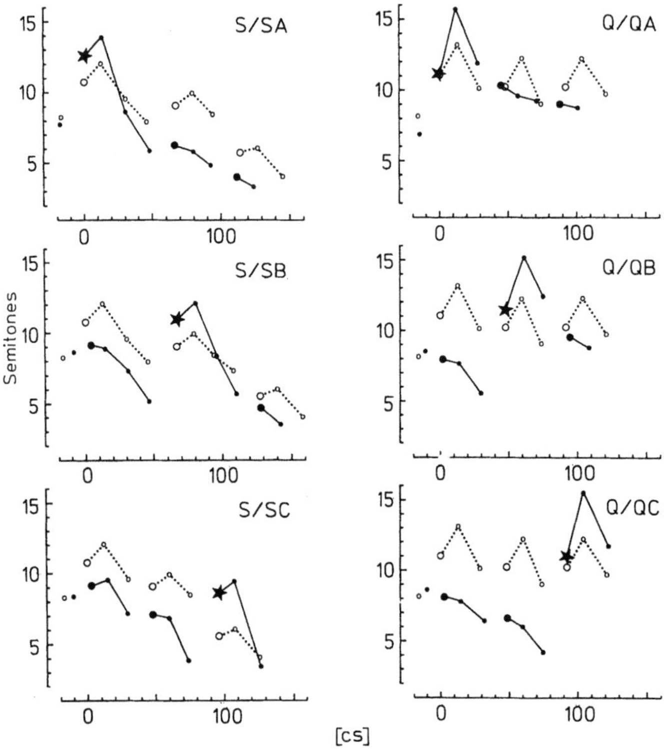

Fig. 3.1. Stylized tracings of the course of fundamental frequency (mean of means over 6 readings by each of 2 subjects and 10 readings by 1 subject) in statements (S – left) and questions (Q – right): prosodically neutral (circles and dotted lines) and with emphasis for contrast (stars, dots, and solid lines) initially (A – top), medially (B – middle), and finally (C – bottom) in the utterance. Stars denote the emphasized syllable, large circles the stressed syllables, and small circles unstressed syllables. Zero on the logarithmic frequency scale corresponds to 100 Hz (see also Thorsen [1980 a])

happens after it [von Essen, 1956; Isačenko and Schädlich, 1970]; word accent differences are not deleted in non-focal position in Swedish [Bruce, 1977]. Furthermore, as I hinted above, some instances of sentence accent presumably fall under the category of "contrast", but the semantic/pragmatic distinction (to the extent that it can be made) and the phonetic differences (if any) between contrast and other types of prominence are rarely made explicit, with one notable exception: From Jones [1960, §§ 1049–1059] we learn that when contrast emphasis is superposed upon a sentence accent in utterance-final position, its F_0 movement is even more elaborate. With contrast emphasis on some earlier word, this F_0 movement is moved back to that word's stressed syllable, and succeeding syllables "have the intonation of unstressed syllables" [Jones, § 1050]. Now, his readers can conclude either that the sentence accent is deleted due to the contrast emphasis earlier in the utterance, or that it is moved back to coincide with the contrast emphasis. The decision is an arbitrary one, because whichever we choose, the fact remains that according to Jones's description, we can distinguish phonetically between a neutral sentence accent and contrast emphasis in utterance-final position but not elsewhere.

The fact that non-final contrastive and non-contrastive sentence accents may not be distinguished phonetically, and that the manifestation of emphasis for contrast in Standard Danish has certain features in common with such non-final sentence accents does not, of course, affect my argument that Standard Danish lacks an obligatory sentence accent, because the test case remains: in pragmatically neutral utterances no special prominence is attached to the last (or any other) stressed syllable.

According to Bruce [personal communication], Southern Swedish also lacks an obligatory sentence accent, so Standard Danish is not a completely isolated case among the Nordic languages. Furthermore, Dutch also seems to defy the sentence accent analysis, but in a different manner: One *or more* of the lexical stresses (but not necessarily all of them) will be manifested as pitch accents, and these pitch accents are apparently all equally prominent ['t Hart and Collier, 1979]. The pragmatic rules which govern the assignment of pitch accents in Dutch are the object of a study by Terken [1980a].

I am of course not blind to the fact that in spontaneous speech in Danish we certainly get uneven stress distributions, i.e. varying degrees of prominence among the stressed syllables of an utterance, without necessarily evoking the impression of emphasis for contrast. There are several reasons why I do not think this kind of emphasis is a sentence accent phenomenon:

(1) There may be any number of emphasized words within one prosodic phrase.
(2) If I can trust my impression from just listening, such non-contrastive emphases do not lead to a deletion of the neighbouring F_0 deflections, so
(3) their prominence derives from increased duration and probably intensity rather than from a stress reduction in the surroundings. Furthermore, a distinction between sentence accent and other types of prominence (emphasis) is probably also required in spontaneous speech in sentence accent languages [Brown et al., 1980].

I do not know what the general psycholinguistic significance of the missing sentence accent in Danish might be, but I am certain that it is – or should be – an important feature in language teaching, both ways: Danes who omit the sentence accent in, say, English – in styles of speech where it would have been appropriate, for instance reading a scientific paper to an audience – are reported by speakers of English to sound dull, deadpan and uninterested in their own subject. On the other hand, a Swede speaking Danish is likely to be labelled "too emphatic", "affected", etc.

3.3 The Representation of Sentence Intonation

3.3.1 The Tone Sequence Approach

I have previously presented data and arguments to support a view of the composition of Standard Danish intonation in terms of a layered system of simultaneous, non-categorical components [Thorsen, 1979, 1980 a, b, in press]. I have also argued that although the stress group and sentence intonation components are highly interactive on the concrete articulatory and physical level (where the stress group pattern is subject to quantitative variation), there may be a level in the speech production and perception processes where the two can be viewed as invariant entities [Thorsen, 1980 a]. Figure 3.2 illustrates three stages in the superposition process, and in Thorsen [1979] an example is given of the inverse process, i.e. the decomposition of the fundamental frequency course in an utterance into its constituent components. Figure 3.3 summarizes the results of analyses of stress group patterns and intonation contours in short utterances.

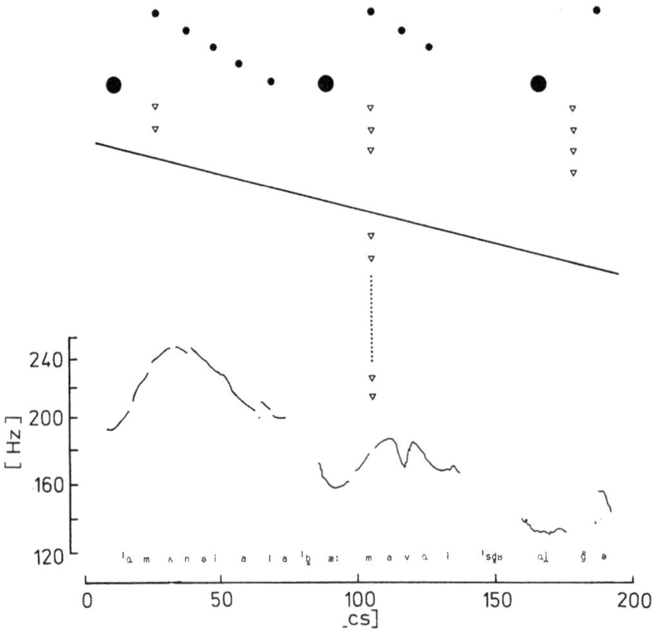

Fig. 3.2. Illustration of the superposition principle in the production of fundamental frequency in an utterance: "Ammerne i Alabama var i strejke" ("The nurses in Alabama were on strike"): Three stress group patterns of different length, superposed upon a declining sentence intonation contour, will – by way of adjustment rules and microprosodic rules – yield the F_0 contour shown (here by a female speaker)

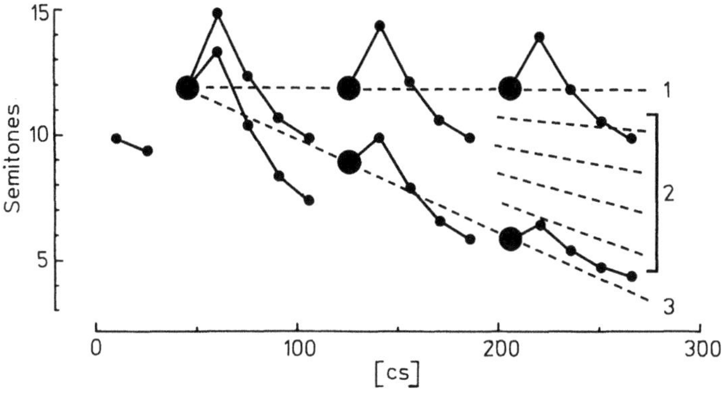

Fig. 3.3. A model for the course of fundamental frequency in short sentences in Standard Copenhagen Danish. (**1**) Syntactically unmarked questions, (**2**) interrogative utterances with word order inversion and/or interrogative particle; non-terminal declarative and interrogative sentences (variable), (**3**) terminal declarative utterances. The large dots represent stressed syllables, the small dots unstressed ones. The solid lines represent the F_0 pattern associated with stress groups, and the dashed lines denote the intonation contours

Pierrehumbert [1980] takes a radically different approach. She suggests that Standard Danish be re-analyzed in a fashion similar to her analysis of English – in terms of a sequence of pitch accents, consisting either of monotonal H or L accents or bitonal combinations of the two (connected with "+" in the transcriptions). In bitonal pitch accents, either of the tones may be the stronger one, lining up with the stressed syllable, leading or lagging behind a weaker tone. Tones aligned with stressed syllables are denoted with a star. Figure 3.4 is an example [Pierrehumbert, personal communication] of the tonal representation of an utterance with different prominence relations between the two stressed syllables.

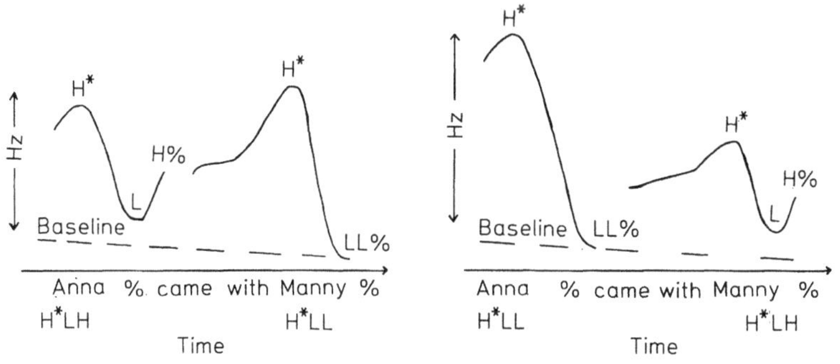

Fig. 3.4. Fundamental frequency tracings and tonal representation of the utterance "Anna came with Manny" with different prominence relations between the two stressed syllables: "Anna" is "backgrounded" in the left-hand tracing; "Manny" is "backgrounded" in the right-hand one [Pierrehumbert, personal communication]

Pierrehumbert observes that in a number of terminal declarative sentences, the degree of overall downdrift is too large to be accounted for by the slight, physiologically determined baseline declination. In order to keep within a framework of categorically different intonational units which are non-interacting and simply follow each other in time, and so maintain a description of intonation in terms of tonal sequences which together make up or, rather, which *are* the intonation of the utterance, Pierrehumbert introduces a *rule of downstep* which will lower a H tone in the context H+L__ and HL+__, independently of the relative prominence (starring) of the tones; compare the top of Fig. 3.5 to the lower tracings. Overall downdrift is a result, then, of contextually determined downsteps which are completely locally governed. The actual scaling of a pitch accent requires no look-ahead and is totally specified by its tonal type, its prominence, the baseline value, and the immediately preceding tone. Further, accents are downstepped through the utterance by a constant factor, smaller than unity, relative to the preceding H, which creates asymptotically declining contours.

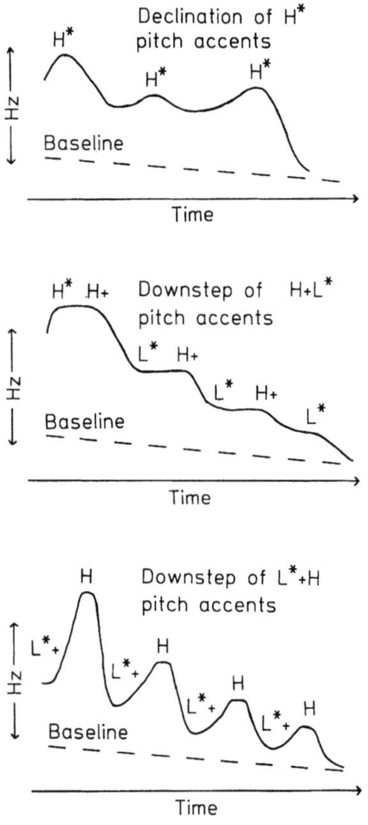

Fig. 3.5. Three examples of fundamental frequency contours and their tonal representation [Pierrehumbert, personal communication]

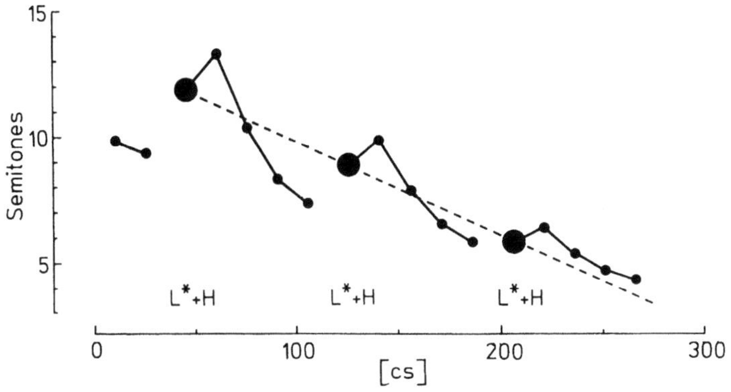

Fig. 3.6. Model for the course of fundamental frequency in a short terminal declarative utterance in Standard Danish, and its tonal representation, as suggested by Pierrehumbert [1980] (See also the caption to Fig. 3.3)

Confronted with the Danish data, particularly the low-plus-high-falling stress group pattern, Pierrehumbert faces what looks on the surface like an obstacle. In her system, Standard Danish has a bitonal L*+H accent (Fig. 3.6). If tritonal accents are to be avoided, there is no room for a L after the H in the Standard Danish pitch accent; i.e. there is no context for the downstep rule. Pierrehumbert circumvents this problem by positing a downstep rule which applies in declaratives in Danish and which downsteps L* in relation to the preceding H. She bases this on an observation from my model (Fig. 3.3) that the L*+H interval is constant, whereas the HL* interval varies with degree of overall downdrift; in other words, the L*+H interval is implemented in a constant way, whereas a L* dissimilates from the preceding H to varying degrees in different types of utterances. With a rule which downsteps L* in relation to the preceding H by a factor *k* which varies according to the status of the sentence, all Danish sentence intonation contours are the product of locally determined pitch accent scaling. (Pierrehumbert's first observation, that the L*+H interval is constant, is not accurate. It may be true at a more abstract level of production and perception, but not at the level where the actual physical scaling of F_0 is taking place, and that is the level Pierrehumbert is dealing with where the downstep rule is concerned).

3.3.2 Problems with the Tone Sequence Approach in Standard Danish

(i) A general point of difficulty with the Tonal Sequence approach is that it seems to operate with an open scale of degree of prominence (cf. Pierrehumbert [1980], Chap. 3). It demands of the speaker and listener (and analyst) that they can reliably and consistently produce and identify a large number of prominence degrees. If such an ability to identify highly varied prominences cannot be assumed with all speakers and hearers, the Tone Sequence theory faces a serious obstacle.

(ii) Pierrehumbert's downstep rule cannot account for the way a sequence of two successive stressed syllables (i.e. with no intervening unstressed syllables) are scaled in, say, a terminal declarative, let alone a whole utterance like *Per så Lis* ['b̥ʰeʌ̯ 'sɔˀ 'lis] ("Per saw Lis"), which is just as downdrifting/downstepping as utterances with unstressed material between the stressed syllables. I do not think that introducing an abstract or underlying H after the L* can solve the problem, since I do not see how an abstract tone can be a factor in the actual computation of the scaling of a succeeding tone. This difficulty with the downstep rule would disappear if L* could be downstepped in relation to the previous L*. That would be much better motivated phonetically and would also remove the difficulty created by the variation in the magnitude of the rise from stressed to first post-tonic (the L*+H interval). However, I suppose that since the previous L* is not always *immediately* preceding, it is unacceptable as a general context for the downstep rule to apply in, because it will disrupt the strict locality principle. This illuminates a fundamental problem with the Tone Sequence approach: Unstressed syllables after the first post-tonic (H) must get their F_0 course by a phonetic fill-in rule. That is in itself reasonable enough, also in the light of my own speculations that stress group patterns may be invariant entities at a higher level in production than the concrete articulatory one. However, in order to scale the succeeding L* stressed syllable, these intermediate unstressed syllables must be disregarded; in other words, the speaker must employ some kind of phonetic "look-back" mechanism to locate the H which – together with other factors – determines the actual frequency value of the next L*. The only reason why this phonetic mechanism might not look back one more syllable to the preceding stressed one (i.e. L*) is the constraint imposed by the tonal representation and the locality principle. That is to say, the distinction between those features of a speaker's output which are relevant to his ongoing production and those which are not is *not* independently established, but is a consequence of the theoretical framework itself. In this particular case it will prevent a phonetically more plausible downstep rule (L* is downstepped after and scaled in relation to L* rather than H), and one which would also reflect the fact that the stressed syllables are the relevant ones in a listener's identification of intonation contours [Thorsen, 1980b].

Ladd [in press, and this volume] modifies Pierrehumbert's theory while staying within its general framework. He finds her account of (Danish) sentence intonation contours unconvincing and introduces a downstep *feature,* which gives separate representation to the overall downward slope, treating it as an independently selectable phonological phenomenon without giving up the advantages of tonal sequence phonetic specification. It seems to me that this has obvious advantages over the downstep rule, and it also solves the problem of environment mentioned above. Ladd suggests that the downstep

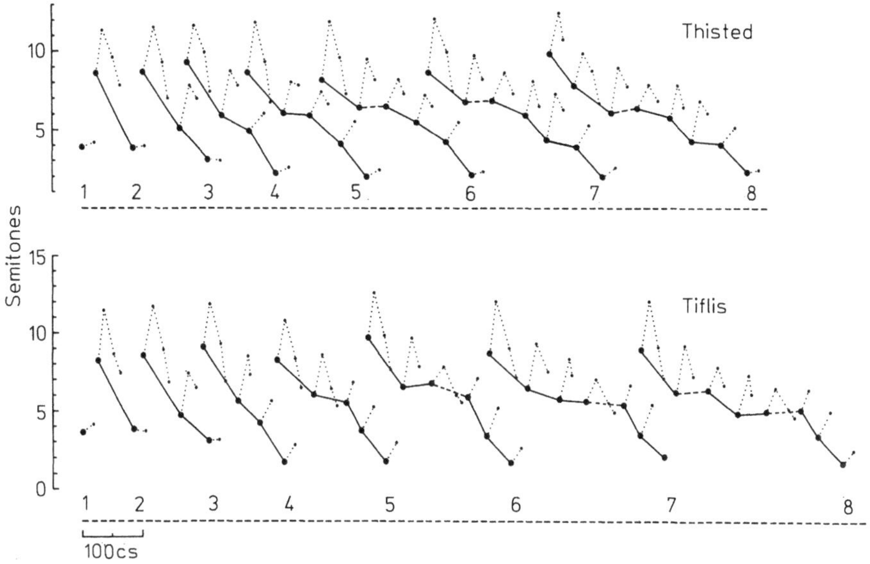

Fig. 3.7. Intonation contours (solid/dashed lines) and stress group patterns (dotted lines) in 2 sets of terminal declarative sentences containing from 1 to 8 stress groups. Average over 4 subjects (mean of means over 6 recordings). Large dots represent stressed syllables, small dots unstressed ones. The dashed lines indicate boundaries between prosodic phrases (see also Thorsen [in press]). Zero on the logarithmic frequency scale corresponds to 100 Hz

feature be carried over to the analysis of English, and together with some other modifications he achieves a rather less abstract tonal representation.

Independently of the choice of either a downstep rule and its environment or a downstep feature, there are, however, still certain points which make the Tone Sequence representation descriptively inadequate for Standard Danish:

(iii) If an intonation contour (or phrase contour) is *not* asymptotically declining, and if this is *not* due to uneven prominence of the stressed syllables, then the downstep factor or feature must of necessity take different values as it creeps along the utterance. Figure 3.7 offers evidence for non-asymptotically declining phrase and intonation contours (see also Thorsen [in press]).

(iv) Consider utterances of varying length: In the conceptually simplest case where range is constant over utterances of different length and where consequently the slope of the intonation contour (the downstep factor) is inversely proportional to the length of the utterance it spans, the downstep must be factored differently in short and long utterances, something which presumably demands a certain amount of look-ahead for its computation. In more complicated cases where range increases with utterance length, but not linearly so, the demands on the computation of the manifestation of the downstep are even more stringent [Thorsen, in press].

(v) Apart from their descriptive adequacy, the two representations may be evaluated in terms of their ability to reflect speech production and perception processes, to the extent that such processes are known, or at least hypothesized. The Tone Sequence theory explicitly requires no look-ahead mechanism, but we know that speakers do look ahead [Perkell, 1980], also in the planning and execution of intonational phenomena [Bruce, 1981; Lehiste, 1975]. This fact is mirrored in the global representation of sentence intonation in the layered system. The stressed syllables (the starred tones) in the Tone Sequence theory have no special role or status in connection with the determination of overall downdrift (downstep) – compare the formulation of the downstep rule above – but at least in Standard Danish the stressed syllables seem to serve as anchor points in the identification of intonation contours, and this fact is also reflected in the layered system.

To conclude, if the downstep rule or feature is to capture the variability and complexity of intonation contours like those presented in Fig. 3.7, it must be a scalar and continuous feature which will often have to take different values within one and the same phrase or sentence. In other words, it must have at least some of the properties it presumably was intended to rid intonation analyses of. It seems to me that the Standard Danish data are more easily accommodated by a descriptive system or theory that requires a look-ahead mechanism and allows its components to be parametric, simultaneous and physically interacting, than by a theory whose components are sequential and categorical and generated without look-ahead from left to right.

4. Peak Features and Overall Slope

D. Robert Ladd

With 8 Figures

4.1 Introduction

The goal of this contribution is to show how the peak-feature model of intonational phonology [Ladd, in press] can be applied to issues of overall slope and declination dealt with elsewhere in this volume, in particular by Gårding, Thorsen, and Vaissière. Unlike these other works, which treat overall slope in terms of grid lines or abstract contour shapes, the approach taken here sees overall slope, in effect, as a by-product of the linguistic specification of accent peaks. This specification is in terms of features that interact to determine peak scaling.

The peak-feature model was developed as part of the intonational meaning project in Giessen, and it reflects the project's experimental bias and concern with pragmatics. For experimental purposes, we want to be able to create, through digital resynthesis, sets of stimuli with systematic differences in their intonation contours which we can then use in judgment studies of pragmatic or attitudinal effects. This obviously requires a painstakingly explicit description of phonetic detail. For purposes of describing intonational meaning, on the other hand, we want to be able to relate pragmatic differences and similarities between contours to systematic differences and similarities in their form. This obviously requires some understanding of the linguistic organization of intonation, that is, some idea of which differences count and which do not. This dual requirement places severe (but, I hope, ultimately useful) restrictions on what we take "systematic difference" to mean in this context. It means that we must not describe intonational form solely in acoustic or psychophysical terms (which would in some sense be adequate for classifying stimuli; see e.g. the work of Takefuta [1975] or de Pijper [1979]), but must take into account higher levels of linguistic organization as well. The most important motivation for the peak-feature model has thus been the attempt to build both phonetic explicitness and functional relevance into the description of intonational form.

4.2 The Peak-Feature Model

4.2.1 Background

Two current general approaches to intonational phonology have made good progress toward explicit description of phonetic detail. We might call these the "contour interaction" (CI) theory and the "tone sequence" (TS) theory. The first is best exemplified in recent work by Thorsen [1978, 1979, 1980 b] and Vaissière [this volume]. They assume that the pitch contour of an utterance is specified by a number of separate components which generate, for prosodic domains of various sizes, pitch configurations that are superimposed or overlaid on one another; a single output contour results from the interaction of the several contours specified by the various components. It seems fairly clear that certain aspects of this model must be incorporated into any complete description, in particular the notion of a "microprosodic" component that specifies details like the intrinsic pitch of vowels, perturbations of F_0 due to stop consonants, etc. [cf. Thorsen, 1980 b; Hirst, this volume].

What is controversial about the contour interaction view as presented by Thorsen or Vaissière is the idea that *accent* and *sentence intonation* are two separate layers or components, with the intonation being specified as an overall line, and the accents being specified as local movements deviating from that line. Against this view, the position of the "tone sequence" theory (represented by Pierrehumbert [1980] and by 't Hart and his colleagues ['t Hart and Cohen, 1973; 't Hart and Collier, 1975]) is that the pitch movements associated with accented syllables are themselves what make up sentence intonation. In this theory, there is no layer or component of intonation separate from accent: intonation *consists of* a sequence of accents, or, to put it more generally, a sequence of tonal elements.

Theoretical problems with both these views are discussed in detail in Ladd [in press]. For our purposes here, we may summarize by saying that both theories appear forced to choose between phonetic explicitness and functional relevance. In a contour-interaction model, functional generalizations are readily expressed in statements about the overall shape or slope of sentence contours [e.g. Thorsen, 1978]. However, as I have argued elsewhere [Ladd, 1981], this process of generalization irons out phonetic detail that may be relevant. Once such detail is acknowledged (as it is in Thorsen [1980 c]), the number of different sentence-intonation types multiplies rapidly, and there is no obvious basis for extracting further explicit generalizations. As for the tone sequence model, in its application by both Pierrehumbert and 't Hart, functional considerations are more or less excluded a priori ['t Hart and Collier, 1978; Pierrehumbert, 1980, p. 59]. Phonetic detail is generated by an inventory of several categorically different accent types strung together in large numbers of possible combinations, and differences of overall slope are generated by the cumbersome method of positing different accent types to make up contours with different slopes (e.g. in Pierrehumbert's system a sequence of H^*+L^- accents has a steeper overall slope than a sequence of otherwise similar H^* accents).

Some of the difficulties with these two theories are avoided in the compromise model proposed by Bruce and Gårding [1978; Gårding, this volume]. In their system, as in the TS theory, contours are generated by a succession of tones (high and low), but the actual height of the highs and lows is determined by reference lines, which, like the CI theory's sentence intonation contours, can vary in their overall shape and slope. While in my opinion this compromise model will ultimately face the same difficulties as the CI model because of this continued reliance on overall contours, it nevertheless represents an important advance, because in other respects it provides the basis for a systematic phonetic taxonomy of pitch that meets the requirements of phonetic explicitness and functional relevance. In developing the peak-feature model I have taken my lead from some of their insights, and in the following subsection I present the notion of features as a formalization of some of their analyses.

4.2.2 Some Details

The two most important analytical insights of Bruce and Gårding's system are (1) abstracting away from the syllable and (2) positing categorical differences of peak height[1]. The first is illustrated in Fig. 4.1, which is based on

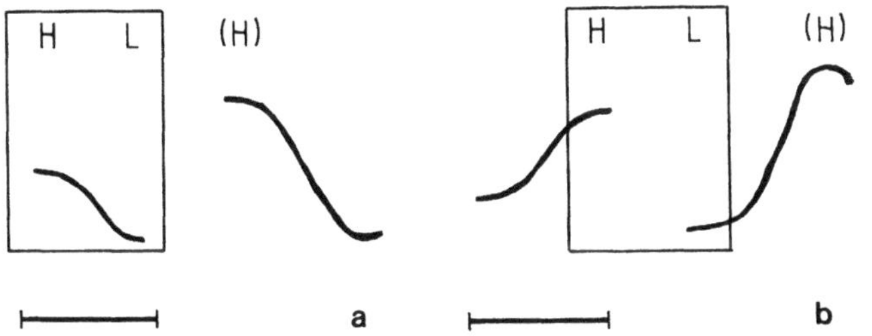

Fig. 4.1. Pitch configurations of Accent 2 in two-syllable words with focus accent in (**a**) the Stockholm and (**b**) the Göteborg dialects of Swedish [Bruce and Gårding, 1978]. The horizontal bar below the contour shows the limits in time of the accented vowel. The boxed part of the contour illustrates Bruce and Gårding's analysis of the difference between the two dialects, in terms of a common high-low (H L) or falling configuration. The parenthesized high following the accent is a sentence-level intonation feature

[1] These two notions are not, of course, entirely original with Bruce and Gårding. The syllable-independent analysis of Scandinavian pitch accents has been in the wind since Rischel [1960] or even Haugen and Joos [1952], while the idea of categorical differences of pitch height in intonation languages goes back to American levels analysts such as Pike [1945] and Trager and Smith [1951]. Nonetheless, Bruce and Gårding are to my knowledge the first to combine these in a single explicit model.

data from Bruce and Gårding [1978] and shows idealized contours of "Accent 2" in two different Swedish dialects. If we attempt to anchor our phonetic description of these accent contours to the limits in time of the accented syllable itself – the horizontal bars – then we are forced to say that Accent 2 is a "rise" in the Göteborg dialect and a "fall" in the Stockholm dialect. This of course makes the Swedish pitch accents seem very mysterious, since it is hard to imagine what might be considered invariant or even vaguely similar in the accent realizations of these two dialects. However, if we look at the accent configuration as defined by its own peaks and valleys, which is what Bruce and Gårding do, then we can factor out, so to speak, an invariant configuration, namely fall, and a variable dimension of "peak alignment", i.e. the timing of the peak with respect to the accented syllable. Here we have the elements of the feature analysis. The fall is a sequence HL, and the alignment is an n-ary phonetic feature on the H. Moreover, as Bruce and Gårding note, differences of alignment not only characterize the phonetics of different dialects, but also signal the distinction between Accent 1 and Accent 2 in any given dialect: the peak of Accent 2 is always reached later than the peak of Accent 1. This means we can speak of a binary phonological feature, say [$^{\pm}$delayed peak], whose exact phonetic values are specified differently from dialect to dialect. In short, we can reduce the differences among the various dialects to variation along a single phonetic dimension, and can make the useful phonological abstractions Accent 1 and Accent 2 with a full understanding of which phonetic detail we are abstracting away from, and why.

More important for the main topic of this paper is the notion of categorical differences of peak height, which is implicit in Bruce and Gårding's use of reference lines. One such line defines the phonetic value for something like "normal high" (their "topline"), while another defines the value of "overhigh" (their "focal line"). In a declarative contour in the Central Swedish dialect, the H associated with the nuclear or focus accent is scaled overhigh, while the other Hs in the utterance are scaled normal high (Fig. 4.2). In the

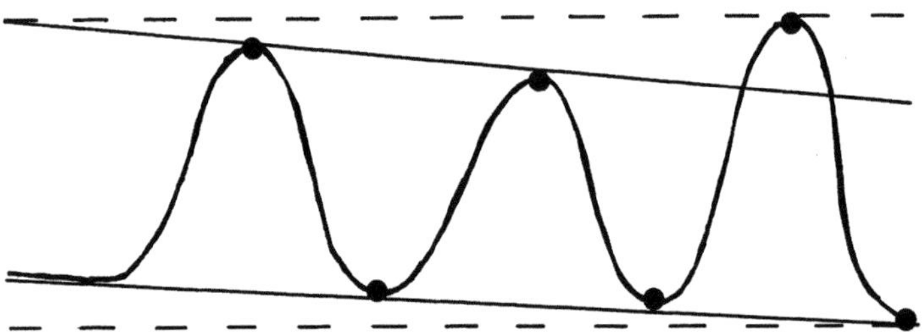

Fig. 4.2. Schema for a sequence of two nonnuclear and one nuclear high-low configuration in Central Swedish [Bruce and Gårding, 1978]. The non nuclear Hs and Ls are scaled on the solid lines; the nuclear H and L are scaled on the more widely spaced dashed lines. Syllable alignment with the peaks is not shown

framework of the feature description, this phenomenon could be interpreted as a feature distinction [±raised peak], which would apply to the nuclear H. Similarly, Gårding's observations [this volume] that the "topline" and "baseline" are higher in questions than in statements might be represented in terms of a feature distinction [±raised range], which would apply to all the tones in the sequence. On any given tone, all the features relevant to height (raised peak, raised range, and probably several others) would then interact to determine the actual phonetic value of the tone along the "vertical" frequency scale.

This formulation is obviously sketchy and somewhat speculative, but it will suffice to give the reader an idea of the type of description I have in mind with the peak-feature model. On the basis of this sketch, it may seem that I have produced the worst sort of "notational variant", simply putting Bruce and Gårding's straightforward descriptive apparatus into more fashionable terms. The rest of the paper is devoted to showing that, on the contrary, the feature model permits more insightful descriptions and makes testably different predictions. The specific questions I will discuss all have to do with the general problem of overall slope.

4.3 Cross-Classification and Functional Relatedness

The most fundamental purpose of any feature analysis is to permit cross-classification of linguistic entities, i.e. simultaneous representation of both their similarities and differences. Equipped with just the two features [±delayed peak] and [±raised peak], we can distinguish four functionally distinct accent patterns in the Central Swedish data presented by Bruce and Gårding[2]. Accent 2 is distinguished from Accent 1 by being [+delayed peak]; nuclear or focal accents are distinguished from nonnuclear ones by being [+raised peak]. These descriptive statements are true independently of one another; that is, we can ignore the difference between Accent 1 and Accent 2 in order to make generalizations about nuclear accents, and by the same token we can ignore differences between nuclear and nonnuclear accents to generalize about the difference between Accent 1 and Accent 2. This can be summarized in the 2 × 2 feature matrix shown in Fig. 4.3.

The descriptive usefulness of such cross-classification can be better appreciated if we take a more complex example. Consider three possible ways of saying the string of digits *02139* (the MIT postal code). A "neutral

[2] The discussion here is limited to Central Swedish data because the other dialects present complications which would detract from the simplicity of the argument. Mutatis mutandis, comparable descriptive statements can be made about the other dialects as well.

	− Raised peak	+ Raised peak
− Delayed peak	Nonnuclear Accent 1	Nuclear Accent 1
+ Delayed peak	Nonnuclear Accent 2	Nuclear Accent 2

Fig. 4.3. Feature analysis of high-low accents in Central Swedish

declarative" way of saying this would be with all the peaks at roughly the same height[3]:

0 2 1 3 9 (4.1)

With a focus on the final digit *9*, to distinguish the sequence from, say, the Harvard postal code 02138, the string might well be uttered as follows:

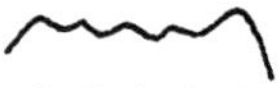

0 2 1 3 9 (4.2)

Finally, in the following we have an intonation suitable for answering someone who has just asked for the tenth time if you're absolutely sure that the MIT postal code isn't 02912:

0 2 1 3 9 (4.3)

With a feature analysis, it is possible to treat all three of these utterances as having similar and functionally related intonation contours (i.e. falling, declarative) without ignoring the differences. The common denominator is that all are a series of H accents followed by a final HL. In (4.1), no further specification is needed. In (4.2), with narrow focus on *9*, we can posit an added feature of raised peak on the nuclear HL, as in the Swedish data discussed above. In (4.3), we have raised peak on the initial H, and a series of "downsteps" on the remaining accents in the sequence. (The notion of a

[3] Like the Swedish data in Figs. 4.1–4.3, the intonation contours in these examples are highly idealized to simplify the exposition. Note that the peaks in Example (1) are at the same height if we allow for the effects of declination in the sense discussed further below.

downstep feature is based on Pierrehumbert [1980] and is discussed further in Ladd [in press].)

The feature analysis thus permits us to ignore differences of phonetic detail in a systematic way, a way that embodies the claim that the intonation contours in (4.1–3) belong to a natural class. The representation H . . . HL shared by these three contours is not simply part of the phonetic notation, but an attempt to represent the fact that the three contours have something in common. It factors out the "declarative" part of the intonation and leaves various other aspects (like "narrow focus" or any affective nuances) unspecified. No such simple description of the functional similarities and differences appears possible in a model that involves overall tunes or reference lines. In an analysis of that sort, all three contours would be seen as different, and there would be no clear basis for calling them similar as well as different. For example, Vaissière [this volume] states that speakers are able to control the rate of declination to signal linguistic distinctions, and presumably the difference between (4.1) and (4.3) would in her view be a matter of rate of declination. Thorsen's view appears to be similar; Bruce and Gårding would presumably specify these contours with different toplines, which also amounts to the same thing. As for the difference between (4.2) and the others, most CI analyses treat "neutral" utterances as a separate descriptive category from ones involving "emphasis" or "contrast" [e.g. Thorsen, this volume], which would presumably make it difficult to relate (4.2) to (4.1) and (4.3).

4.4 Peak Scaling and Declination

The fact that the feature model seems to permit more insightful functional descriptions of intonation is certainly of interest, but that is too subjective a criterion to make a convincing case for the model's usefulness. In order to provide a clear demonstration of its empirical potential, we must compare the feature model's claims with those of the CI model on data that involve less intuitive judgment and evaluation and more instrumental or experimental evidence. This section discusses the different views of "declination" that are implied by the two models and presents some preliminary evidence in support of the feature model account[4].

"Declination" refers to the overall downward tendency of F_0 throughout the course of an utterance. This has been discussed and noticed in a variety of languages and by a variety of investigators, and has been explained and analyzed in various ways. In what follows I will consider two major

[4] These data are taken from work in progress and their preliminary nature will be obvious. A more definitive study is in preparation.

approaches, which we might call the regression-line view and the zero-line view.

The regression-line view is that declination is manifested by abstract lines drawn through selected points (usually accent peaks) in actual contours. This approach is taken by a considerable number of investigators, e.g. Thorsen [especially 1979], Vaissière [this volume], Cooper and Sorensen [1981], and, in effect, Bruce and Gårding. (A variant of this approach draws lines through *un*accented syllables, e.g. Brown [this volume].) The regression lines can obviously differ considerably as to slope (steeper or shallower) and shape (straight, asymptotic, irregular, etc.); this makes the regression-line view of declination particularly consistent with the CI model of intonation in general, since both imply that sentence intonation involves a choice among a number of overall shapes and/or slopes, global contours with which the local accent contours interact or to which the local accent contours are fitted.

The zero-line view of declination involves the stronger and more abstract hypothesis that declination is more or less invariant, a gradual (probably physiologically determined) lowering of an abstract zero value from which accent peaks are scaled. This view has been put forth most clearly by Pierrehumbert [1980], but there are also elements of it in Collier [1975 b] and Maeda [1976][5]. One of the consequences of this view for interpreting F_0 data is that the "topline" of an utterance is taken to be, at best, an indirect reflection of declination, since the height of the peaks through which the topline is drawn is determined by a number of factors, not just the declining value of "zero". This view of declination is thus quite consistent with the feature model of intonation in general: in both, the question of "overall slope" is not really a matter of overall slope at all, but of *scaling of accent peaks*.

These two different views of declination associated with the two models of intonational phonology make claims about the shape of contours that are clear enough and different enough to permit empirical investigation. I will discuss three specific cases:

(1) the presence or absence of declination in intonationally marked questions;

(2) the possibility of distinguishing declination from "downstep" – i.e. two different mechanisms that lower F_0 during an utterance;

(3) the possibility of explaining apparently random variation around regression lines.

[5] Maeda and Pierrehumbert assume that the absolute pitch drop is the same regardless of utterance length, yielding steeper slopes in shorter utterances. In 't Hart [1979] a formula is given which increases the pitch drop somewhat with the length of the utterance but which still yields steeper slopes in shorter utterances. In this discussion I have assumed a constant drop. In any case, it must be kept in mind that "invariant" declination does not mean that the slope is invariant.

4.4.1 Declination in Questions

The CI view assumes that the “rise” found in intonationally marked questions in many languages is due to an overall upward-sloping (or at least not downward-sloping) “sentence intonation” contour on which the accent perturbations ride. This implies that declination is somehow “turned off” by the speaker in such utterances (a position adopted explicitly by Vaissière in this volume). The strongest version of the opposing view is that question rises result from linguistic choices of tones or pitch accents, and it predicts that the gradual automatic drop in the value of zero during the course of the utterance should be unaffected by these choices.

Evidence for the second position can be seen in Fig. 4.4, which shows the extracted F_0 curves of three questions taken from our corpus of quasi-natural speech in Giessen[6]. In each case it can be seen how the pitch rises from the accented syllable or syllables and then gradually declines from its peak until it

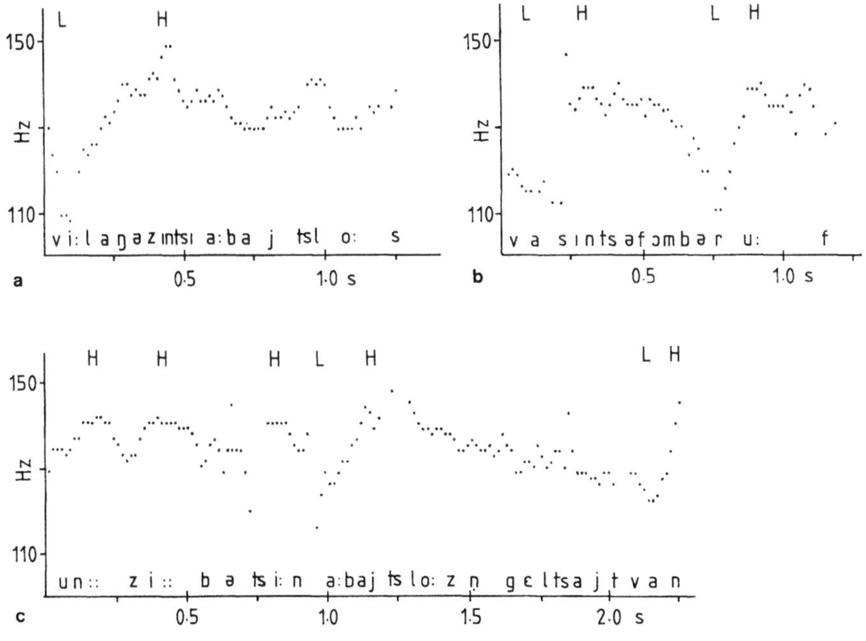

Fig. 4.4. Three rising question contours, exhibiting declination in addition to rising accents. Likely locations of highs and lows are indicated. The texts are as follows: (**a**) *Wie lange sind Sie arbeitslos?* (main accent on *Wie*); (**b**) *Was sind Sie von Beruf?* (**c**) *Und – Sie – beziehen Arbeitslosengeld seit wann?* (hesitation on first two words)

[6] The corpus consists of interviews between German social workers and actors playing clients, recorded during a social skills workshop at the University of Giessen. Further details can be found in Scherer and Scherer [1980].

steps down to the next accented syllable (or in the case of Part (a), until it reaches the end of the utterance). These rises can be analyzed as LH pitch accents; the unaccented syllables following the H are unspecified for pitch and simply remain high, but the actual F_0 value of "high" gradually declines because of the automatic decline of the zero-line.

4.4.2 Declination and Downstep

Differences in overall contour slope of the sort seen in examples (4.1) and (4.3) above are attributed in the CI model to differently specified sentence intonation contours or declination rates. The feature model account of the difference is ostensibly much more complicated: it assumes, as we have seen, that the rate of declination is essentially invariant and that the different overall slopes are due to different specifications of pitch height above the declining value of zero. Specifically, cases like (4.1) would be treated as having equal peaks, so that the slope of the "topline" would roughly parallel that of the zero-line, while in cases like (4.3) each peak would be downstepped relative to the preceding peak, yielding a topline noticeably steeper than the zero-line[7]. In order to justify the second view, we need to show that it is possible to distinguish the effects of declination and downstep on F_0 given an appropriate context.

Such a context should be provided by sentences in which several syllables intervene between accented syllables: we should be able to observe a gradual decline on the unaccented stretch that is distinct from the abrupt downstep at the accented syllable. Schematically, we should expect contours like the following (where *s* indicates unaccented syllables and *S* accented syllables):

s s S s s s s s S s s s s s s S s s s s (4.4)

This appears to be the pattern observed in Fig. 4.5, also taken from our corpus. Here the downsteps on *Ar–* and *–komm–* are clearly identifiable, but the overall downward trend of the stretches between the accented steps is equally clear. Obviously, this downward trend is much less steep than the regression line drawn through the accented syllables, which suggests that the two cannot be accounted for in the same way; the declination-plus-downstep analysis provides two independently specified mechanisms to account for the two phenomena. It is not clear how the CI model would explain the details of F_0 in this example.

---

[7] This approach is essentially that of Pierrehumbert [1980].

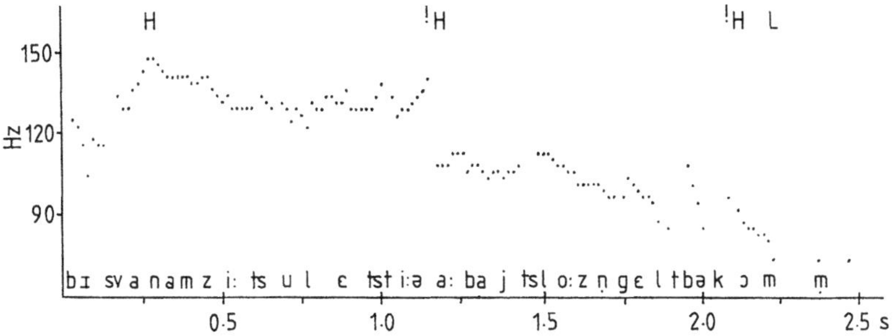

Fig. 4.5. A question contour exhibiting downstep as well as declination. Likely locations of highs and lows are marked; !H indicates downstepped high. The text is as follows: *Bis wann haben Sie zuletzt Ihr Arbeitslosengeld bekommen?*

4.4.3 Deviations from Regression Lines

The third type of evidence in favor of a zero-line approach to declination is that it appears to provide principled explanations of data that a regression-line approach must treat as random variation. This is best seen in the following data from Cooper and Sorensen [1981, pp. 76–85]. They measured F_0 at the four major accent peaks in short, medium, and long sentences of the following general form:

The house on the corner was built by his grandfather. (4.5 a)
The house on the corner, you realize, was built by his grandfather. (4.5 b)
The house on the corner, you must have realized by now, was built by his grandfather. (4.5 c)

The average toplines they obtained for the three groups of sentences, together with the toplines predicted by their formula, are shown in Fig. 4.6. In all cases the value of the third peak is some 5% less than the predicted value; this error is acknowledged but unexplained [Cooper and Sorensen, 1981, p. 85].

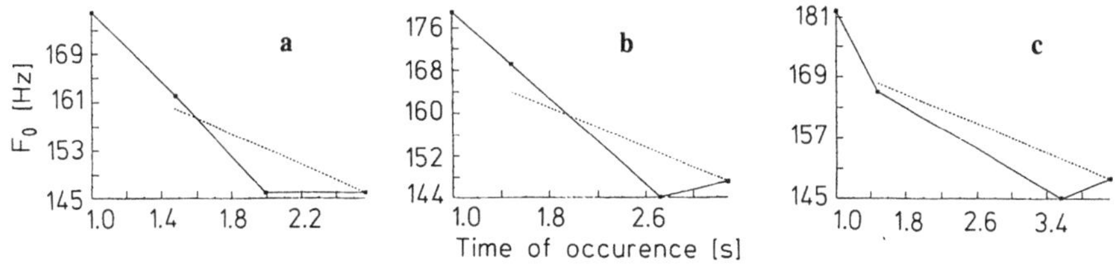

Fig. 4.6. Predicted (dotted) and obtained (solid) toplines in (**a**) short, (**b**) medium, and (**c**) long utterances [Cooper and Sorensen, 1981]. (See also Fig. 4.8.)

To be sure, the "error" could be reduced substantially by abandoning the assumption that the topline should be straight. An irregular line that dipped for peak 3 and then rose a little for peak 4 would represent a substantial improvement. Not all CI models assume that the lines connecting accents should be straight or otherwise generable by a simple function: Thorsen [1980 c] admits the existence of irregular intonation lines, though she is at a loss to account for them or fit them into her overall view of sentence intonation. Yet even a line that dipped for peak 3 would still leave the details of the relation between peak 3 and peak 4 as unexplained variation. That is, it would provide no account of the fact that peak 4 is the same as peak 3 in the short sentences and higher in the longer ones.

In a zero-line model, by contrast, this fact arises naturally as a consequence of the model. Assume that peak 3 and peak 4 are scaled fixed amounts above a declining zero-line – say x and $x + y$ units respectively – in all three sentences. Since the slope of the zero-line is shallower in longer utterances (see Footnote 5), peak 4 has to be absolutely higher in order to maintain the same relative height. This is shown schematically in Fig. 4.7.

In a general way, it is clear that the behavior of peaks 3 and 4 in Cooper and Sorensen's data fits the situation shown in Fig. 4.7. In Fig. 4.8, the model is applied to their data in more detail: the three toplines from Fig. 4.6 are shown again, this time together with toplines predicted by computing peak height above zero-lines. In all three sets of sentences, the four predicted peaks have been scaled 4.5, 3.6, 1.9, and 2.5 semitones, respectively, above zero-lines that decline 2 semitones[8]. The fit between predicted and actual values is obviously quite good. To be sure, the predicted values are in some sense ad hoc, chosen to give a good fit; it is clearly a task for future research on the peak-feature model to determine more explicitly the linguistic basis on which peaks are scaled. Nevertheless, the very fact that it is possible to account for the data in all three sets of sentences with the same zero-line declination and the same peak values suggests that there may be some validity to the general approach, since nothing in Cooper and Sorensen's model would lead us to expect a systematic relationship of this sort between peaks in different versions of the same sentence. The more basic point of the example is also clear: the regression line model accepts fairly substantial deviations from predicted values as part of the inherent variability in the data, while the

[8] The zero-lines begin and end slightly higher, the longer the sentence; the higher beginning is consistent with the results of Cooper and Sorensen [1981, Chap. 2], or of O'Shaughnessy [1976], while the higher end is consistent with the findings of Thorsen [1980 b]. The amount of declination, 2 semitones, is equivalent to 15 or 20 Hz at this range, which is a rough estimate consistent with Maeda's physiological data (see also Pierrehumbert [1980, Chap. 3]). The utterances are assumed to begin 0.4 s before the first peak and end 0.5 s after the last peak; it is unfortunately not possible to do more than estimate, since the only time data given by Cooper and Sorensen are normalized times of occurrences of peaks. The data have been converted to semitones following the usage of Thorsen [1980 a] and 't Hart [1979]; this, of course, is also a matter for empirical investigation.

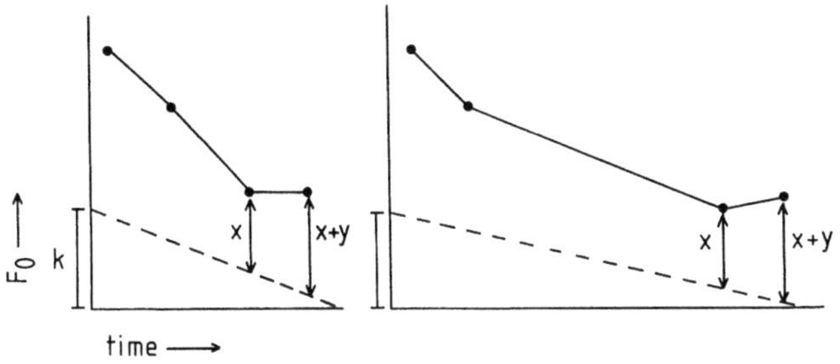

Fig. 4.7. Schematic drawing showing how different slope of topline between peaks 3 and 4 in short and long contours (solid lines) could represent a constant relationship to baseline (dashed lines), assuming that the baseline declines by a constant amount regardless of utterance length

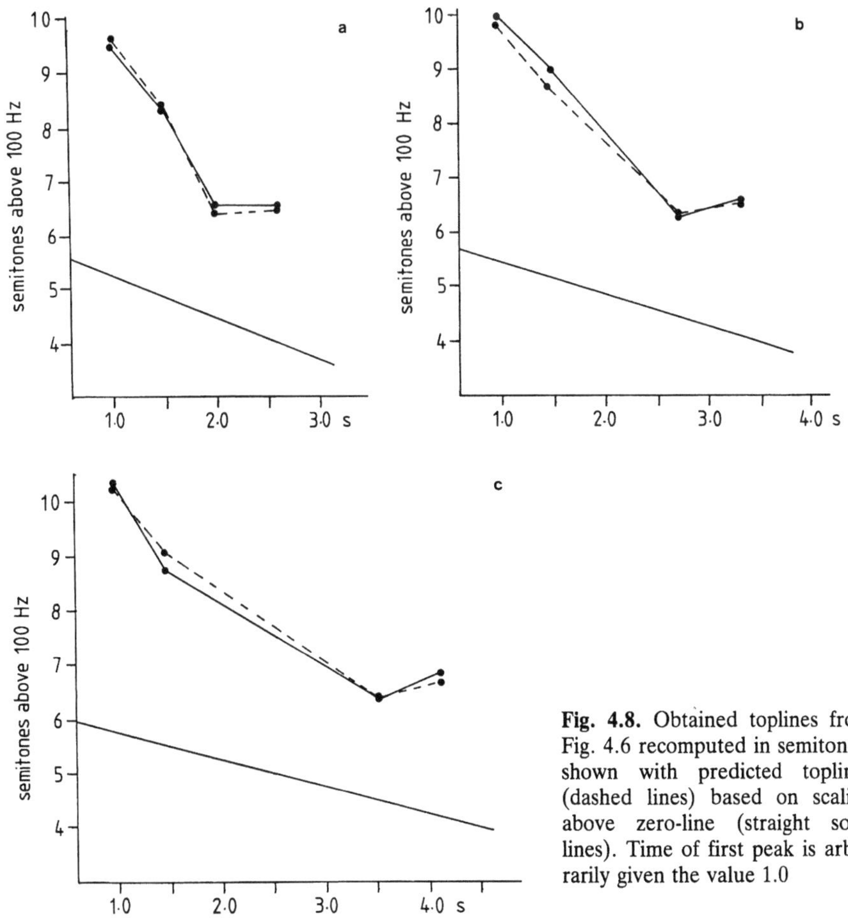

Fig. 4.8. Obtained toplines from Fig. 4.6 recomputed in semitones, shown with predicted toplines (dashed lines) based on scaling above zero-line (straight solid lines). Time of first peak is arbitrarily given the value 1.0

zero-line model in which peaks are scaled individually appears to permit – and require – more precise predictions. To the extent that these predictions can be made in terms of categories expressed as peak features, there will be additional evidence of the feature model's usefulness in representing intonation.

4.5 Summary

I have briefly outlined the peak-feature representation of intonation and compared it with other current models on the basis of data involving overall slope. The feature model appears superior to models that specify overall tunes in its ability to express generalizations about intonational function. I have also shown how the feature model favors a zero-line view of declination, in which declination is treated as a gradual invariant drop in the value of an abstract zero from which individual accent peaks are scaled. This model of declination accounts for phonetic data better than models involving overall tunes or regression lines. Taken together, the functional data and the phonetic data suggest that the peak-feature model represents a promising approach to intonational phonology.

Acknowledgement. The work reported here was supported by the Deutsche Forschungsgemeinschaft. I thank my two collaborators on the intonational meaning project, Klaus Scherer and Kim Silverman, for helpful criticism and discussion.

5. Language-Independent Prosodic Features

Jacqueline Vaissière

With 2 Figures

5.1 Introduction

The purpose of this contribution is to investigate the similarities in form and function of prosody among diverse languages. All speakers, regardless of their specific language, are equipped with the same production and perception apparatus, and consequently have the same capabilities and must face the same physiological constraints. Such similarities should be reflected in the acoustic production of any speaker. The first specific aim of this contribution is to review a number of striking acoustic similarities in the suprasegmental aspects of neutral sentences in different languages, together with possible physiological explanations for them.

Since the global function of speech (i.e. communication) is language independent, the type of information conveyed by the speech signal should not vary greatly among languages. The second purpose here is thus to compare the linguistic functions assigned in several languages to similar suprasegmental phenomena, such as lengthening, fundamental frequency rises and falls, and intensity peaks.

Although the use of suprasegmental variation for paralinguistic functions (such as the expression of emotion) is also of prime importance in everyday conversation, the scope of this contribution is restricted to the strictly linguistic functions of prosody.

5.2 Language-Independent Similarities

5.2.1 Pauses

Acoustic analysis of speech production shows that speakers insert a large number of pauses while talking. Studies have indicated that the ratio of articulated sequences to total speaking time does not differ significantly from one language to another: at least, Grosjean and Deschamps [1975] have found no difference for such a ratio between native speakers of English and of French in radio interviews (about 83% for both languages).

The first language-independent primary function assigned to pauses is that of grouping. Speakers tend to draw breath at the end of large conceptual units such as sentences and clauses. Pauses between sentences tend to be longer than pauses within sentences (Goldman-Eisler [1972] for English; Grosjean and Deschamps [1973] and Duez [1982] for French). The use of pauses as a major boundary marker between and within sentences seems to be similar across those languages for which there are available data. Black et al. [1966] found no significant differences in the use or location of pauses by speakers of English, Hindi, Japanese and Spanish. However, the length of pauses and their distribution per number of syllables may differ slightly from one language to another: Grosjean and Deschamps [1975] found that native speakers of French tend to pause less often but for a longer time than native speakers of English.

There also exists another type of pause, generally not located at grammatical junctures, the so-called hesitation pause. The duration and frequency of both types of pauses seem to depend on a larger number of factors, such as speech rate, speech mode, age and experience of the speaker, emotional state, and the pursuit of stylistic effects. At fast speech rates, pauses within sentences tend to be suppressed [Grosjean and Collins, 1979]; pauses are more frequent in descriptions than in interviews [Grosjean and Deschamps, 1973]; in political speeches, pause distribution tends to be more in agreement with grammatical structure than it is in interviews [Duez, 1982]. German and English adolescents make more ungrammatical pauses than adults while reading [O'Connell and Kowal, 1972]. Advanced students make longer and more judicious pauses in reading than inexperienced students [Clemmer et al., 1979]. In the sixth repetition of a spontaneous description, subjects made only half as many pauses as in the first repetition [Goldman-Eisler, 1968], and made fewer and shorter pauses when reading unusual stories as compared to ordinary ones [O'Connell et al., 1969]. Variations in semantic context can produce changes in the number of unfilled pauses in a situation in which syntactic and other variations are minimized, for both German and English speakers [O'Connell and Kowal, 1972]. Pauses also depend on the emotional state of the speaker [Fairbanks and Hoaglin, 1941], and are often used before words with low transition probability [Cowan and Bloch, 1948; Maclay and Osgood, 1959]. All such factors seem to be largely independent of language.

Speech is produced by modulating the airflow from the lungs. As a consequence, it is necessary for all speakers to insert a certain number of respiratory pauses while speaking. Respiratory pauses, which represent only a part of all pauses, can be considered physiological necessities. The constraints due to the production apparatus explain the basic similarities in pausing between languages. The probability of occurrence of a pause during continuous speech seems to depend on the amount of residual air in the lungs [Fujisaki and Sudo, 1971]. It has been shown by Grosjean and Deschamps [1975] that breathing during speech exerts an influence at similar time intervals independently of the language spoken (here English and French). The most striking similarity between languages is that breathing (respiratory pausing) occurs at grammatical junctures. At fast rates, the physiological need to breathe is the sole determinant of pausing [Grosjean and Collins, 1979]. Fodor et al. [1974] have suggested that breathing is dependent on syntax: speakers will only breathe when allowed to do so, as it were, by the constituent structure of the utterance.

5.2.2 Fundamental Frequency Features

Figure 5.1 summarizes the general properties of F_0 curves in simple sentences pronounced in a single breath-group (that is, without pauses), as observed in a number of languages. We will review in turn the declination tendency of the F_0 curve, the resetting of the baseline, the control of F_0 range, and F_0 rising versus F_0 falling movements.

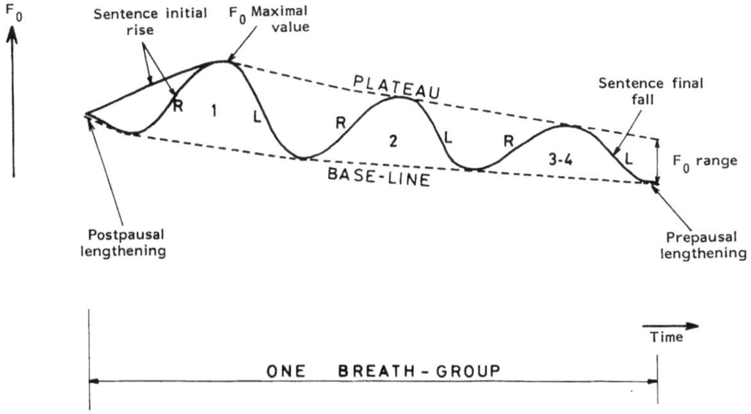

Fig. 5.1. General properties of F_0 contours observed in unmarked sentences in a number of languages. The common tendencies are the following: (i) a tendency for F_0 values to fluctuate between two abstract lines: the *plateau* and the *baseline*, which delimit the speaker's F_0 range; (ii) a tendency for the F_0 *range* to diminish as a function of time; (iii) a tendency to start sentences with a large *sentence-initial rise* in F_0, located on one of the first syllables, or spread over the first few syllables; (iv) a tendency to repeat a succession of F_0 rises (R) and lowerings (L): a pair of opposing movements indicates a prosodic word (see text); (v) a tendency for the *maximal value* of F_0 to be located on the first prosodic word of the sentence; (vi) a tendency to lengthen the duration of the last syllable at the end of the breath-group (*prepausal lengthening*), and of the first phoneme at the beginning of the sentence (*postpausal lengthening*)

a) The Declination Tendency

In relatively long stretches of continuous speech, there is a global tendency for the F_0 curve to decline with time, despite successive local rises and falls

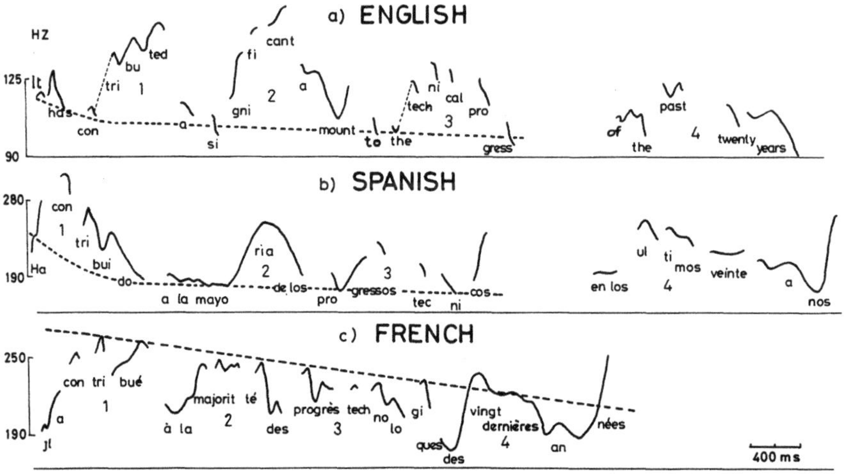

Fig. 5.2 a–c. F_0 contours for the same clause translated into three languages and pronounced by native speakers: (**a**) "It has contributed a significant amount to the technical progress of the past twenty years, . . ."; (**b**) "Ha contribuido a la mayoria de los progressos tecnicos en los ultimos veinte anos, . . ."; (**c**) "Il a contribué à la majorité des progrès technologiques des vingt dernières années, . . .". The pauses and the rise-fall pairs in F_0 combine to divide the three clauses into four prosodic words (numbered 1 to 4)

(for Dutch, Cohen and 't Hart [1967]; for English, Maeda [1976]; for Italian, Magno-Caldognetto et al. [1978]; for Japanese, Fujisaki et al. [1979]; for tone languages, Hombert [1974]). Figure 5.2, which shows the F_0 contours of the same utterance translated into three different languages and spoken by a native speaker of each language, illustrates this tendency.

Several physiological explanations have been proposed. The F_0 declination could be mainly due to the decline in transglottal pressure (proposed by Lieberman [1967] and partially refuted by Hixon et al. [1971], to a trachea pull [Maeda, 1976, p. 220ff.], or to a laziness principle (the rise supposedly being harder to produce than the falls [Ohala and Ewan, 1973]). The different explanations primarily invoke the production apparatus and are compatible with one another. The declination tendency has also been proposed to be a by-product of the perceptual system [Pierrehumbert, 1979]. The rate of declination has been found to vary with a number of factors, but no cross-linguistic differences have yet been proved.

Declination is a general tendency easily detected from a visual inspection of relatively long stretches of F_0 curve, but calculation of the exact rate of declination is a difficult task. For some speakers, the tendency seems to be adequately described by a constant rate of declination ("the declination line"), while for others, there is a tendency for F_0 to decrease more rapidly in the first part of the sentence, and then to decrease less rapidly or even stay flat (exponential declination). The slope of declination is not entirely physiologically determined, but is partially controlled by the speaker. It has been shown that the speaker often adjusts the rate of declination with the length of the sentence, at least in isolated sentences: the shorter the sentence, the more rapid the declination. When the sentences are part of a text, the correlation between rate of declination and length of utterance is less obvious; in spontaneous speech, many sentences do not even display such a tendency. Furthermore, the declination tendency has been shown to be often suppressed in interrogative sentences (both in spontaneous speech and in read sentences).

It may be suggested that the rate of declination contributes (together with other features) to the perception of a sentence spoken in the declarative mode as an acoustic whole; its presence versus its absence contributes to the contrast between declarative and interrogative sentences (for French, Contini and Boé [1975]; for Danish, Thorsen [1980 a]; for Russian, Svetozarova [1975]; for a tone language, Ho [1977]; for a number of other languages, Lieberman [1967]).

We may conclude that as for pauses, the natural tendency for F_0 to decline has been integrated into the linguistic code, in the form of controlling or entirely suppressing the F_0 decline. However, its role is probably of minor importance in spontaneous speech, even though it may contribute to naturalness.

b) Resetting of the Baseline

A direct consequence of this tendency for F_0 to decline is that F_0 values tend to be lower near the end of breath-groups than at the beginning. After a pause, they are generally reset. We may suppose that resetting of F_0 values after a respiratory pause is a natural physiological tendency, although the relation between resetting and breathing is not entirely clear.

Together with the declination tendency and pauses, the resetting of F_0 may contribute to characterizing sentences and clauses as acoustic entities. It has been shown (by Maeda [1976] for English; by Carlson and Granström [1973] for Swedish; and by Fujisaki et al. [1979] for Japanese) that resetting of the baseline can be simulated between two phrases, even in the absence of a pause. The resetting is used as a boundary marker, and the degree of resetting indicates the importance of the boundary.

A stepping down of the baseline at major syntactic boundaries has been reported in a few languages [e.g. Ford, 1975]. To decide on the general use of baseline resetting, either upward or downward, acoustic data for a larger number of languages, particularly tone languages, must be compared and described within a common framework.

The physiological basis of the relations among pause, breathing, declination and resetting is difficult to establish, since speakers may pause without breathing, or reset the baseline without pausing; moreover, as noted above, declination has also been seen as a by-product of the perceptual system. Despite these difficulties, the considerable cross-linguistic similarities suggest a physiological explanation.

c) Normal Frequency Range and the Control of Range

The range of F_0 variation generally narrows as a function of time (the range is often widened at the very end of the breath-group with the realization of an optional continuation rise). In general, the local F_0 maxima and minima decrease from the beginning to the end in simple declarative sentences; the local maximal values tend to decrease more rapidly than the local minimal values (Fig. 5.1).

This diminution of F_0 range seems to be a natural physiological tendency, since it has been attested in simple sentences in a large number of languages. Most of the physiological explanations for such a phenomenon relate to the production apparatus. It has been shown in singing that an upward pitch change takes longer than a downward pitch change, suggesting that it is easier to lower pitch than to raise it [Ohala and Ewan, 1973]. (Note that a difference in difficulty between the two opposite gestures has also been observed in deaf people [Salles, 1980].) As a consequence of this tendency, the increase in amplitude of successive F_0 rises does not entirely compensate for the decrease in falls, resulting in a narrowing of the F_0 range. A perceptual explanation is sometimes advanced: it has been argued that the speaker tends to keep the same number of octaves above the baseline.

The largest rise (and consequently the highest F_0 value) is generally located in one of the first three or four syllables of the utterance (that is, within the first lexical word). This seems to be common to the languages that have been studied so far. However, speakers are able to counteract this supposed natural tendency to a certain extent, creating a local widening or narrowing of the expected range. Such a strategy has been attested in several languages for marking boundaries or emphasis in complex sentences.

Another observation is that speakers have a tendency to be at a low fundamental frequency, generally the lowest in the sentence, when they cease voicing. The final fall is often accompanied by a lower intensity and by a lengthening of the final elements, which all help to indicate to the listener that the utterance is ending. This tendency is sometimes interpreted as an (early) relaxation gesture [Lindblom, 1968]. Again, however, speakers may counteract this tendency in order to convey linguistic information. The terminal contour of the last syllable is the most important feature in distinguishing simple declarative statements from yes/no questions in a number of languages (for Danish, Thorsen [1980b]; for Italian, Magno-Caldognetto et al. [1978]; for French, Contini and Boé [1975]; for Japanese, Nishinuma [1979]; for Romanian, Dascălu [1979]). When used for marking interrogation, a rising terminal contour is often accompanied by a suppression of the declination tendency. It has been noted, however, that this effect is language specific: in some languages interrogatives can be characterized by a falling intonation, and in others a rising contour may be typical for declaratives. Even in yes/no questions, where the interrogative mode is indicated by the syntax, and in wh-questions, a rising contour is optional.

The terminal contour inside a declarative sentence is also often used to contrast between terminal juncture (at the end of the sentence) and non-terminal juncture (before a non-final pause). (See Delattre [1963] for a comparative study of German, French, Spanish and English; Svetozarova [1975] for Russian and Nishinuma [1979] for Japanese.) We have noticed that in non-terminal juncture, in contrast to yes/no questions, the declination tendency is maintained. The presence versus absence of the declination tendency may contribute to the perceptual differentiation between interrogation rise and continuation rise in a number of languages. Cross-linguistic perceptual experiments are needed to test this hypothesis.

d) Rising Versus Falling F_0 Movements

Perhaps because a F_0 rise is mainly realized by tensing the vocal folds, and a F_0 fall by relaxing the tension, an upward change of F_0 between two successive vowels seems to be associated with the notion of beginning and a downward movement with the notion of end. This tendency seems to be reflected particularly by the manner in which lexical entities are acoustically combined in spoken sentences to form a single acoustic entity. Such an acoustic entity is related to the notions of "stress group" as defined by Thorsen [1980 a] for Danish, the "hat-pattern" in Dutch [Collier and 't Hart, 1975] and in English [Maeda, 1974], the "rhythmic unit" [Lehiste, 1975] or the "prosodic word" in French [Vaissière, 1974, 1975], and the "syntagma" in Russian [Kozhevnikov and Chistovich, 1965; Svetozarova, 1975]. The number of prosodic words for one sentence is not fixed; i.e. it is not entirely determined by grammatical structure. It depends on several factors, such as rate of speech and a speaker's individual choice. However, in carefully spoken sentences, where the speaker tends to stress every meaningful word (generally the lexical words as opposed to the function words), the number of prosodic words (comprising essentially a successive rise and fall) tends to be equal to the number of lexical words, or one fewer (the fall on the next to last lexical word may be suppressed so that a larger fall, the sentence-final fall, is superimposed upon the last word (for French, Vaissière [1975]; for Russian, Svetozarova [1975]). Such an observation is more easily made with sentences composed of multisyllabic words, or sentences where one or more grammatical words occur between the monosyllabic words, that is, where it is possible to establish perceptual contrast between successive stressed and unstressed syllables. This characteristic appears to be language independent.

Many languages seem to use suprasegmental variation to achieve acoustic grouping of semantically related words such as a noun and an adjective which precedes or follows it. (See the above references and, for Chinese, Zee and Maddieson [1979]). This is done firstly by reducing the number of F_0 movements that would be attached to the words if they were spoken separately: the regrouped words tend to receive the F_0 pattern of a single word. Secondly, grouping requires that there be no more than one rise and no more than one fall in the phrasal group; if both occur, the rise has to precede the fall. Figure 5.2 illustrates some regroupings comprising a F_0 rise in the first lexical word and a F_0 fall in the last word: “to the technical progress”, “of the past twenty years” (English), “des progrès technologiques” (French), “en los ultimos veinte anos” (Spanish).

5.2.3 Durational Features and Intensity Phenomena

a) Final Lengthening

There is a tendency to lengthen the final elements in an utterance, particularly the last vowel, before a pause (see illustrations of prepausal lengthening for French, English, German and Spanish [Delattre, 1966]; for Italian [Marcel, 1971]; for Russian [Zlatoustova, 1954]; for Swedish [Lindblom, 1968]). However, there seem to be languages in which there is little (if any) final lengthening, e.g. Finnish [Lehiste, 1965], Estonian [Lehiste, 1965] and Japanese [Han, 1961].

It has been suggested that slowing down at the end is a natural tendency characterizing all motor sequences or planning units: similar patterns have been found in music, in birdsongs, in insect chirps and of course in speech [Cooper, 1976]. The lengthening of final elements in words, phrases and sentences in speech is an intriguing phenomenon. It seems that lengthening, like F_0 fall, is associated mainly with the notion of termination.

Lindblom [1968] refers to Öhman’s [1967] model of word and sentence intonation, in which there is a “general relaxation of speech gestures toward the end of the utterances”. Making the assumption that the same amount of physiological energy is expended for each syllable (excluding stress variations), Lindblom concludes that the increased duration of final syllables compensates for the lower intensity per unit of time. However, Oller and Smith [1977], who approached the issue from a developmental perspective by studying the linguistic production of infants, concluded that final-syllable lengthening constitutes a learned behavior, since infant speech does not display such a tendency.

Lengthening a final element without a pause helps to mark the end of a word or phrase (for English, Delattre [1966] and Oller [1973]; for French, Delattre [1966]; for German, Zingle [1974]; for Russian, Zlatoustova [1954]; for Spanish, Delattre [1966]; for Swedish, Lindblom [1968]). It has been observed that phrase-final lengthening has a larger range than word-final lengthening. Cooper [1976] has noted a related tendency in music. Variations in relative lengthening of different elements, if perceived, contribute to the recovery of syntactic structure by the listener: the greater the lengthening, the more important the syntactic break (see the experiment by O'Malley et al. [1973] on the recovery of the syntactic structure of algebraic expressions).

In addition to lengthening of the very last syllable in a word or phrase, there may be three other kinds of lengthening phenomena: first, lengthening of the last stressed syllable in a phrase, even if it is followed by one or more unstressed syllables; second, lengthening of the entire last word in a phrase [Umeda and Quinn, 1981]; and third, lengthening of the last sentence in a read paragraph [Lehiste, 1975].

The bulk of final-syllable lengthening (which may be used as a right boundary marker) seems to be carried mainly by the vowels [Oller, 1973]. By contrast, word-initial consonant lengthening has generally been attested as a left boundary marker, a function associated with rising F_0 (for English, Lehiste [1960]; for Italian, Nespor [1977]; for Swedish, Carlson and Granström [1973]).

b) Other Lengthening Phenomena

While lengthening of the final element seems to have essentially a demarcative function, lengthening in a non-final context is largely used for emphasis marking. The lengthening of a single syllable is commonly used to mark emphasis and contrastive stress [Coker and Umeda, 1973]; deceleration in speaking rate is a strategy that may be used by some speakers to emphasize a word, an entire sentence or a clause [Kloker, 1975].

5.2.4 Intensity and the Syllable

Prosodic phenomena such as tone, stress and accent are customarily considered qualities of a syllable. The syllable is the minimal unit for F_0, duration and intensity contrasts. Even the mora, which may be more adequate than the syllable for describing the suprasegmental system of languages like Ancient Greek [Garde, 1968], is defined relative to the syllable. Despite language-specific rules for grouping phonemes into syllables, there is a universal tendency to bring the structure of the syllable as close as possible to the ideal described by de Saussure [1916] and Jespersen [1933].

A common definition of the syllable is that it corresponds to a peak of intensity. Intensity has a demarcative function at different levels. First, the presence versus absence of intensity over a period of time signals pauses and divides speech from non-speech sequences. Second, during articulated sequences, the gross variations in intensity and the location of local peaks and valleys chunk the continuum into syllable-like segments. The peaks generally correspond to syllable nuclei (vowels) and the valleys to consonants.

The linguistic function of the relative intensity of syllables in continuous speech is not easy to determine. As mentioned by Sorin [1981], intensity has been less well studied in the past than the two other prosodic parameters, F_0 and duration. There is very little information on the specific contribution of intensity to perception. Intensity seems to share some of the functions of both F_0 variation and durational variation. The interrelation between intensity and fundamental frequency has a physiological basis. Intensity and fundamental frequency are regulated by the same mechanisms (increase in pulmonary effort and in subglottal pressure, tension of the vocal folds, etc.), and, as may be expected, higher fundamental frequency is generally correlated with an increase in intensity (see an illustration of this tendency in Fig. 5.2). Similarly, the F_0 decrease at the end of sentences is matched by an intensity decrease. Higher intensity of a syllable is also often correlated with longer duration. An increase in global articulatory effort to produce perceptual prominence via intensity may also require additional time, and consequently may result in a lengthening of the underlying phoneme. For example, emphasis is often marked by both an increase in intensity and a lengthening of the underlying segments.

Generally speaking, higher F_0, higher intensity and longer duration naturally tend to characterize the same segments in a sentence. Such a tendency is particularly obvious in emphasis and prominence marking. However, different languages have counteracted this natural tendency to a greater or lesser degree in establishing their specific suprasegmental code. We will see in the next section that a particular relationship between intensity, F_0 and duration may in fact be a characteristic of a language.

5.3 Prosodic Differences Among Languages

The broad similarities among languages should not mask the considerable variations indicated by cross-linguistic comparison of prosodic systems [Ladd, 1981]. There seem to be three basic principles which guide the comparison and description of such differences.

5.3.1 Differences in Timing

In English [Maeda, 1974], Italian [Magno-Caldognetto et al., 1978] and Dutch [Cohen and 't Hart, 1967], pitch movements (either upward or downward) are mainly limited to the so-called stressed syllables (with an optional continuation rise on the last unstressed syllable). Together with higher intensity and longer duration, they help to assign prominence to stressed syllables relative to the surrounding syllables. The position of the stressed syllable in a word is largely predictable from the phonological and morphological structure of the word. In French, F_0 movements occur in relation to word boundaries: a rise is superimposed on the beginning of the word, and the word-final syllable carries a large variety of accentual contrasts (large rise, rise, level, fall, large fall). In the former three languages, F_0, duration and intensity variations are mainly "timed" with the stressed syllables, while in French, they are bound to the first and last syllable of the word.

Moreover, the exact position of rises and falls in so-called stress languages varies from one language to another. The movements are not always simultaneous with the stressed syllable. In Danish, for example, a stressed syllable in neutral speech is relatively low, and is followed by a high-falling tail on the following unstressed syllable, a pattern which is not common in English. The stressed syllable(s) may be associated in Danish with two different accents, which are mainly distinguished by the presence or absence of a kind of glottal stop (called stod), rather than by a F_0 movement [Gårding, 1977].

The direction of the pitch change, either rising or falling, is determined by language-specific factors. We have noted that in French, the importance of the juncture following the word determines the "accentual pattern" (rising, falling or level) of the word-final syllable. The F_0 pattern superimposed on a word in that language is mainly determined by the sentence context. In Swedish, by contrast, one syllable in the word may receive primary stress (the position of which is predictable from the phonological and morphological structure of the word), and one syllable with primary stress may carry one of two typical accents [Gårding, 1977]. Syntactic structure also seems to play a role in Swedish in determining the actual F_0 pattern realized upon a word embedded in a sentence, but to a lesser degree than in French. In a tone language [Pike, 1945], where there is a lexically significant tone on each syllable, the syntactic structure may be expected to play an even smaller role in determining the fundamental frequency contour than in Swedish.

As a consequence, word boundaries, morphological and phonological word structure, and juncture phenomena may all contribute to the determination of F_0 contour in such languages. The relative importance of each factor is language specific.

Differences in the timing of the same movements are also one way of explaining dialectal varieties in a given language. The best demonstration is probably given by the work of Bruce and Gårding on Swedish [Bruce and Gårding, 1978; Gårding, 1975 a]: F_0 peaks may occur more or less early in the stressed syllable, or even in the post-stressed syllable, depending on the dialect spoken (see also Fujisaki and Sugito [1976] for Japanese).

A different timing rule has also been attested for speakers of the same community. It has been shown, for example, that native speakers of French may start a continuation rise at the onset of the syllable, or later, in the vowel [Vaissière, 1975].

5.3.2 Different Orders of Priorities

As the notion of differing orders of priorities is relatively new, we shall illustrate it through examples.

a) Due to the physical constraints imposed by the vocal apparatus producing F_0 curves, and to the limitation of the F_0 range, it is necessary for upward and downward F_0 movements to be produced alternately, so that a large fall is preceded by a rise, and vice versa. In English and in French, a rise may serve different purposes: it may indicate a non-terminal juncture (continuation rise) in both languages, and/or be used to mark a stressed syllable (in English) or the beginning of a word (in French). In a language like English, a lowering of F_0 occurs on the last stressed syllable in a phrase. If the last stressed syllable is the final syllable of that phrase, and if this phrase is followed by a non-final pause, either a fall may occur on that syllable to mark the last stressed syllable of the phrase, or a rise may occur to mark that the phrase is non-terminal: data clearly show that priority is given to the fall (stress marking), and that the fall is followed by an optional continuation rise. In French, the continuation rise has higher priority than the fall, which is a usual cue for a right word boundary inside the phrase. As a consequence of this, the fall is overridden by the continuation rise.

b) Both English and French tend to show lengthening of word-final syllables. In English, the acoustic correlates of word boundaries are only traces: marking of word boundaries competes with lexical stress marking, and priority is given to the latter. For example, the longest syllable in the word is generally the stressed syllable, despite a lengthening of the word-final syllable. In French, the marking of word boundaries is not in competition with stress marking: the lengthening of final syllables is "reinforced". The differences between these two originally related languages may be considered a historically different choice of which feature to assign priority: stress or boundary.

Slight differences in priority may also explain dialectal differences. In French, for example, both right and left word boundaries can be perceptually marked. At a normal rate of conversation, however, it is not possible to put prominence on two successive syllables: the prominence assigned to one syllable is achieved at the expense of the syllables immediately preceding and following. In dissyllabic words, native speakers of Southern French give more perceptual prominence to the word ending, while Parisian speakers tend to give almost equal weight to the two syllables. Some speakers, particularly politicians and radio announcers, tend more frequently to "stress" the word beginnings. In words of more than two syllables, it is possible to contrast the two extreme syllables relative to the intermediate syllable(s): the dialectal and individual differences are smaller.

5.3.3 Different Relationships Between F_0, Duration and Intensity

As was noted in the first part of this contribution, intrinsic fundamental frequency, duration and intensity of individual sounds reflect the nature of the speech mechanism in general, and the same intrinsic tendencies seem to exist across languages [Chen, 1970; House and Fairbanks, 1953; Delattre, 1962]; ceteris paribus, for example, higher vowels tend to be uttered with a higher fundamental frequency. Extrinsic variations, which are not due to segment identity, but rather to lexical stress position, accent pattern, or syntactic boundary position, show correlations between the three parameters which may be more or less intimate. In English, the syllables which bear pitch movements (the stressed syllables) are also usually more intense and longer. In French, the last syllable, which carries a variety of F_0 contours, is also the longest syllable in the word; it is not clear whether it is more intense, particularly in level or falling F_0 contours. In that language speakers tend to give a more equal intensity to sequences of sounds than do speakers of English. The three suprasegmental parameters seem to be less systematically correlated in French than in English. In both languages, a longer duration characterizes the syllables carrying pitch movements. But again, such a correlation between F_0 change and duration may not hold equally for all other languages. It has been shown that in Japanese, accent placement has a smaller effect upon segment duration than it does in English [Hirose et al., 1978; Mitsuya and Sugito, 1978].

It is difficult to evaluate the mutual influence of suprasegmental and segmental aspects of one language. Rising fundamental frequency with a suprasegmental origin, for example, is mainly achieved by a tensing of the vocal folds. But the tense consonants, too, are uttered with more tense vocal folds than the lax consonants. An upward change of pitch seems to be favored by the presence of a tense consonant, in comparison with other environments, e.g. lax consonants or vowels. In French, the F_0 maximum after the word-initial rise is located on the word-initial syllable if the word begins with a voiceless consonant; but it may be delayed to the second syllable if the word-initial phoneme is a vowel or voiced consonant; i.e. the position of the turning point in the F_0 curve is determined by a combination of suprasegmental and segmental features. In English, a F_0 rise (stress-marking rise) results in aspiration of tense consonants beginning the syllable on which it is realized. Both phenomena, rising fundamental frequency and aspiration, correspond to a tensing of the vocal cords, and they are generally concomitant in this language. In French, by contrast, a F_0 rise in the onset of a word may or may not be accompanied by aspiration, depending on whether or not the speaker wants to put emphasis on that word.

Suprasegmental aspects of each syllable affect the segmental characteristics of the nucleus. Delattre's comparative study of vowel reduction phenomena in French, English, German and Spanish [Delattre, 1963] illustrates important cross-linguistic differences.

It is possible that the specific *inter*relations between the three suprasegmental features (F_0, duration and intensity), and the relation between segmental and suprasegmental aspects, rather than each feature taken individually, are the most salient characteristics differentiating between languages, dialects, and individual ways of speaking. If this is true, most of the existing descriptions of prosodic systems (including our own work on French) are incomplete, since they describe only one parameter at a time. Compensatory effects between the three parameters have occasionally been noticed. We have learned from our experiments on automatic speech recognition that with some speakers, the longer duration of final syllables is a reliable cue for automatic segmentation of speech into prosodic word-like units, while with other speakers, F_0 variations are more efficiently used.

5.4 Conclusion

This contribution has been based on the hypothesis that the suprasegmental similarities between languages are most likely due to identical contraints exercised by the production and perception apparatus. Its aim was to show how a number of related and unrelated languages have tended to assign the same primary functions to physiologically related phenomena: the necessity of pausing for breath while speaking is commonly used for grouping, the natural tendency for F_0 values to decline with time serves to characterize neutral declarative sentences, and F_0 movements corresponding to a tensing of the vocal folds (resetting of the baseline and local F_0 rises) are associated with the notion of beginning, while falls are associated with ending. We have mentioned examples of how the natural, physiologically conditioned tendencies are controlled in position and amplitude, suppressed, counteracted or simulated to carry linguistic information which is similar in a number of languages: non-respiratory pauses serve as juncture phenomena, the length of pauses is adjusted according to the importance of the juncture, the rate of F_0 declination is adapted to the length of the sentence, the tendency to decline and to end speaking with a F_0 fall are counteracted to provide a contrast between declarative and interrogative modes, and so on.

On the other hand, considerable cross-linguistic differences exist. We have proposed three basic sources of these: different timing of essentially comparable phenomena, different orders of priority for the realization of comparable conflicts between prosodic phenomena, and different relationships between or different mutual effects of F_0, duration and intensity.

6. Prosodic Structure and the Given/New Distinction

Gillian Brown

6.1 The Given/New Distinction

Over the last 15 years, since Halliday drew the attention of scholars in the West to the Prague School division of information within an information unit into *given* or *new* information, a considerable literature has developed. This literature has now largely obscured the phenomenon to which Halliday [1967 a] sought to draw attention. The aim of this contribution is to reassert Halliday's basic distinction, to outline briefly how it relates to the plethora of other distinctions which have since been made in the literature, and to demonstrate from a limited corpus of data that Halliday's simple dichotomous distinctions must be invoked to account for the range of intonational realisation in that data.

Halliday's preoccupation through a series of articles [1963, 1967 a, b, 1970] was to account for the way in which intonation in British English relates to information structure. His account relates exclusively to information structure in spoken language, and the fundamental categories which signal information structure are phonological (with an auxiliary, but never overriding, syntactic contribution). He identifies the unit of information as the *tone group,* i.e. a unit which is intonationally defined [Halliday, 1967 a, p. 200]. According to his classification the speaker must include in every tone group a chunk of new information, which will be phonologically marked by the tonic pitch movement. The speaker may optionally include one or more chunks of given information, which will not be phonologically marked by pitch prominence. It is important to note that in Halliday's account, the assignment of given/new status to information is determined by the speaker, not by the text. Furthermore, whereas Halliday does find some correlation between clause and tone group in the data on which he bases his analytic framework [1967 a, p. 201], in that there is a tendency for the tone group to be co-extensive with the clause, this is merely a tendency, not a requirement on the speaker. It certainly does not follow that long, often complex, written sentences will contain the type of simple information structure associated with tone groups in spoken language (cf. the problems encountered by Prince [1981] in trying to analyse complex expository prose in terms of an analysis developed for the typically short tone-group structures found in spoken language). In Halliday's terms then, the unit of information, the tone group, is phonologically identified, and the status of information within the tone group is phonologically marked in the focussed domain as *new* if the tonic is present, and as *given* if it is absent. The hearer should be able to determine, simply by listening, how an utterance is chunked into tone groups, where within each tone group the speaker has decided to locate the new information, whether he has decided to include given information, and if so, where.

There are problems with the analytic framework even at this level. Several analysts have found intonational criteria inadequate for the identification of information units in speech, and prefer to work with pause-defined units [Chafe, 1979; Brown et al., 1980; Butterworth, 1980; Deese, 1980], and others are anxious to soften the requirement that there should be only one tonic, or marker of new information, within a tone group [Bolinger, 1970; Stockwell, 1972; Crystal, 1975; Chafe, 1979; Pellowe and Jones, 1979; Brown et al., 1980; Thompson, 1980].

For my purposes here I shall assume that the speaker has available *syntax* and *pausing* as two independent but frequently co-operative markers of information chunking, and that there may be several peaks of prominence within a pause-defined unit, which may mark change of topic, speaker's emphasis, contrast, or the status "new information". (For a more detailed discussion, see Brown et al., [1980].) I shall assume that intonational prominence of this kind generally functions as an attention marker, or generalised "watch-it" marker, for the hearer, whereas the function of low pitch is to mark material which the speaker expects the hearer to expect at this point in the discourse [Chafe, 1970]. Note that I am assuming, with Halliday, that the speaker will judge what it is reasonable to expect the hearer to expect at a particular point.

In Sect. 6.2, I shall give a brief overview of how Halliday's given/new distinction has been picked up, particularly in the psycholinguistic literature, and made to do far more work than it can usefully do. As a result, "given" is frequently extended to include all that an addressee can ever be expected to know from whatever source. This extension, particularly when applied to written language examples, often causes the status "given" to be attributed to forms which would inevitably attract phonological prominence if read aloud.

In Sect. 6.3, I shall give a brief account of a restricted spoken language corpus in which the parameter of source of knowledge was carefully controlled. I shall show how this corpus, analysed in terms of the taxonomy developed by Prince [1981], offers support for Halliday's analysis of information status as realised by intonation.

6.2 Extensions of the Given/New Distinction

The extension of Halliday's categories seems to have been initiated because of the way he characterised the state of knowledge which the speaker attributes to the hearer. Thus "given" information is specified as being treated by the speaker as "recoverable either anaphorically or situationally" [Halliday, 1967 a, p. 211] and "new" information is said to be focal "not in the sense that it cannot have been previously mentioned, although it is often the case that it has not been, but in the sense that the speaker presents it as not being recoverable from the preceding discourse" [1967 a, p. 204]. Whereas this characterisation nicely discriminates between the status of information marked intonationally as "given" or "new" by the speaker, it also embraces a wide range of other phenomena. As Dahl remarks, "The concepts of old and new information are used to explain such phenomena in language as intonation, stress and word order and the use of anaphoric devices" [1976, p. 37].

The psychological notion of givenness has been characterised in a range of descriptions, particularly following a series of influential statements by Clark [Haviland and Clark, 1974; Clark and Clark, 1977; Clark and Haviland, 1977] which makes appeal to salient properties of the discourse or context such as: last-mentioned item in the discourse, mentioned earlier in the discourse, ongoing topic of the discourse, known property of a mentioned item, present in the context, saliently present in the context, general knowledge relating to the topic, background knowledge deriving from shared schemata/scenarios, etc. Perhaps the most expansive version of what is to be taken as given can be found in Sanford and Garrod [1981, p. 114]: "The scenario enables references to individuals to be made in the first instance by a definite noun-phrase, because they are already *given* in the representation. Because they are given, they cause neither comment nor difficulty, and rapid pseudo-anaphoric mapping is possible". This makes it clear that any information deriving from the activated scenario has the status of given, not because the speaker (or writer, since they are dealing with written language) assigns it given status, but because of its status within the scenario which the text activates. It would follow from this view of scenario-determined givenness that a speaker giving a stereotypical account of an incident would do so without marking any new information intonationally (resulting, presumably, in a (near) monotone). It is clear that this view of "givenness" does share features with Halliday's ("treated as recoverable either anaphorically or situationally") but that it has little to do with the way speakers use intonation to mark the status of information in spoken language.

The direction that this development of the extension of the term "given" has taken appears to be as follows. Halliday observes that forms in the speaker's message can be characterised as either being phonologically prominent or not. He examines the distribution and function of these forms and concludes that they are used to mark the status of information as given or new. He characterises given and new in terms of speaker assumptions, and exemplifies the realisation of the categories, necessarily, in terms of orthographically transcribed examples of a type which he hopes will yield the intonation patterns he is concerned with, if they are read aloud. The examples of the category "given" typically contain anaphoric expressions, since anaphoric expressions are very often realised by the speaker as not phonologically prominent. Scholars who are primarily interested in written language then read Halliday's written examples and interpret his distinction in terms of the syntactic forms which he has used in his exemplification. Generalising over these syntactic forms, they then observe further syntactic forms in written language which appear to have similar features, and argue that any information expressed in terms of this set of forms is being treated as given by the speaker/writer. Since the range of characterising expressions has been extended far beyond phonological prominence/non-prominence, the meaning of given has necessarily undergone a dramatic change. The set of formal expressions included under its wider interpretation have little to do with the deployment by the speaker of intonational prominence, and frequently, as I shall show in Sect. 6.3, expressions which in the extended meaning of the term would be classified as given are treated by speakers as new, in Halliday's terms. It is a Canute-like act to try to constrain the extended meaning of a term, but it must be a matter of lively regret that this particular extension deprives us of a straightforward way of referring to the phenomena which Halliday was concerned to describe.

A recent survey [Prince, 1981] develops a taxonomy to make it possible to distinguish the various sources of knowledge which lead the speaker to make use of different forms among the formal "state of knowledge" devices which English permits[1]. She distinguishes between *new, inferable* and *evoked* entities. *New* entities may be either *brand-new* (not assumed to be known by the hearer) or *unused* (known to the hearer but not assumed to be in the hearer's consciousness). *Inferable* entities are introduced when the speaker assumes that the hearer can infer them by reasoning from discourse entities already evoked (e.g. the *driver* is inferable from *a bus*, plus assumed knowledge about buses, i.e. *Buses have drivers* [Prince, 1981, p. 236]). *Evoked* entities are ones which have either already been mentioned in the discourse or are situationally salient. This is a useful attempt at trying to get some handle on the way different assumptions by the speaker about the hearer's state of knowledge lead to the use of different formal expressions in language. However, although Prince analyses an extended transcription of a spoken text, she totally ignores the phenomenon of intonation. She attempts only to account for a range of verbal expressions. How does intonation relate to this taxonomy?

In the following section I shall examine the ways in which a number of speakers performing in a restricted speech context exploited the resources of intonation in indicating to their hearers the status of the information they

[1] A good deal of Prince's discussion of Halliday in this survey is vitiated by a misunderstanding of Halliday's position. Prince writes (and reiterates) "For Halliday ... in an information unit with unmarked focus nothing is given" [1981, p. 226]. However, on the very page she cites (p. 208), Halliday states "An item with unmarked focus may thus be represented as being ambiguous, as having the structure *either given-new or simply new*" [1967 a, my italics].

were transferring. I shall show that despite the greater richness of Prince's taxonomy, we still need to appeal to a simple dichotomous given/new distinction in order to account for the intonational realisations, and to a yet richer set of distinctions to account for the range of syntactic realisations.

6.3 An Experimental Study of Intonation and Information Structure

6.3.1 The Data

The data was produced by 12 pairs of undergraduates talking together[2]. In each pair, A could see a diagram which B could not see, and A was asked to describe the diagram to B in such a way that B could draw it. B was provided with a blank sheet of paper, a black pen and a red pen. This highly restricted data produces some characteristic features which I will comment upon later. From the point of view of the analyst interested in information structure, it offers a satisfying amount of control over the range of sources of knowledge. The analyst knows that this is not a fragment of discourse broken out of some previous experience shared by the two participants. He can record all the speech relevant to the transmission of this information. He can identify the point where each entity is introduced into the discourse for the first time, and he can track through second and subsequent mentions. He can recognise where a *brand-new*[3] entity is introduced into the discourse, as in:

a) draw *a black triangle*
b) draw *a straight line*
c) write *OUT in black*
d) there's *a circle* in the middle

He can identify an *inferable* item introduced into the discourse:

a) it's right through *the middle* (circle)
b) you start at *the edge* (triangle)
c) (triangle) with *the right angle*
d) (triangle) *the corner*

---

[2] The data described here derives from SSRC Project HR 6519, "The Information Structure of Spoken Discourse". I am deeply indebted to George Yule, who worked on this project, for some of the analysis and many of the insights which I draw upon here. (For an extended analysis see Yule [1981 a].) This discussion is limited to entity-referring expressions.
I am also grateful to Anne Anderson for the statistics showing the significance of the figures in Table 6.1.

[3] There are no instances of Prince's *unused* category in this data.

He can track second and subsequent mentions (Prince's *textually evoked* category)

a) underneath *the triangle*
b) draw a red line ... at the end + the right hand-side + of *this line* write the word ON just above *the line*
c) a line ... about two inches + and above *it* write ON

He can also recognise contextual referring expressions (Prince's *situationally evoked* category)

a) in the middle of *the page*
b) *you*'ve got a triangle

It could of course be argued that it is unrealistic to distinguish between the categories *inferable, textually evoked* and *contextually evoked,* particularly with respect to this sort of data. The discourse-initial relevant entities in the context may be only page, pen and interlocutor, but as soon as the hearer has drawn, say, a triangle, which necessarily provides him with a set of edges, corners, angles, lines, etc., the context has changed. The hearer is not simply working in terms of a mental scenario; he actually has a model physically present on the page. We might expect, then, that the speaker would collapse these categories and distinguish only between expressions introducing brand-new entities and the rest. However, as we see in Table 6.1, speakers appear

Table 6.1. Distribution of formal realisations in entity-referring expressions from an experimental study of intonation and information structure (in %)

	New Brand-new	New Inferred	Evoked Context	Evoked Current	Evoked Displaced
(a) *Syntactic form*					
a + properties	77				
a − properties	21				
the + properties		29		3	44
the − properties	2	71	58	21	56
pronoun			9	65	
ellipsis			33	11	
(b) *Phonological prominence*[a]					
+	87	79	2		4
−	13	21	98	100	96

[a] Scores are percentages of realisations which receive a physical form.

to distinguish between these categories. For instance, *all* expressions introducing inferable entities consist of full noun phrases, in most cases accompanied by phonological prominence which is not characteristic of forms referring to the evoked context. It seems reasonable to suggest that whereas the relevant features of the interlocutor and the page are thoroughly determined by the nature of the task, the relevant features of the drawing the hearer produces are not so determined.

6.3.2 Formal Realisations in the Data

Table 6.1 shows the distribution of formal realisations in entity-referring expressions in this data. The types of formal realisation include different syntactic forms as well as variation in phonological prominence. The syntactic forms consist of expressions preceded by the indefinite or definite article with or without some specification of the properties of the entity (e.g. "a line" versus "a small red line") and pronominal realisations (in most cases "it", but also "this" and "you"). The ellipsis line in the table represents omissions, as in the expressions

in the middle ϕ draw a black triangle

you've got a red 5 + on the left-hand side ϕ

Phonological prominence is indicated by high pitch in the Edinburgh Scottish English spoken in most of this corpus. The high pitch occurs as excursions from a rather flat baseline of unstressed syllables. In general, words which are phonologically prominent will be recognised by judges as containing "tonic" or "nuclear" syllables. Phonologically prominent syllables occur more frequently in this data than Halliday's tone-group-based definition would predict. Some of the speakers, however, had Glasgow-type accents, and produced pitch excursions below the baseline of unstressed syllables, excursions which curved up to the baseline. Lack of phonological prominence yields a syllable close to the baseline with little, if any pitch movement. Such non-prominent syllables may still be perceived as stressed. (For a description of Edinburgh Scottish English, including a detailed account of the phenomena mentioned here, see Brown et al. [1980].)[4]

[4] The observed differences in phonological prominence when tested statistically were found to be highly reliable in all cases. Binomial tests show phonological prominence to occur more frequently than non-prominence in mentions of brand-new entities and mentions of new inferred entities ($Z = -11$, $p < 0.001$ and $z = -4.98$, $p < 0.01$ respectively). Phonological non-prominence was found to occur more frequently than prominence in cases where the entities were evoked context, evoked current and evoked displaced ($Z = -10$, $p < 0.01$ for evoked displaced).

Continued on page 74

It is immediately obvious from Table 6.1 that *brand-new* entities are typically introduced by expressions with the indefinite article, usually with one or more specified properties, followed by a noun:

draw *a triangle*
draw *a straight line*
there's *a red letter X*
a small black 5

There are a few occasions where an entity is introduced by a definite referring expression, as in:

write *OUT* in black
write *the word ON*

In 87% of cases there is phonological prominence associated with expressions introducing a brand-new entity (this is quite frequently associated with the specified property rather than with the entity-referring noun, as Yule [1981 a] points out). What is surprising here is the number of expressions which are not associated with phonological prominence. Most of these apparently arise from the same source; although the entity introduced is brand-new, the expression used to introduce it is of a form which has already been used to introduce a previous entity. Consider this extract from a speaker whose stable baseline is around 200 Hz and whose prominent range normally lies above 230 Hz:[5]

```
the corner is on the right of the big black  triangle +
      240–210          210–200           200 200      200

a  right-angled  triangle  like  the black one +
     210  210         200                      200    200
```

[4] Continued from page 73:
The frequencies of occurrence of the various syntactic forms used by speakers were also tested, and significant differences were found in mentions of brand-new entities, where more {a} + properties than {a} – properties descriptions were used ($Z = -8.46$, $p < 0.001$), in mentions of new inferred entities, where more {the} – properties than {the} + properties descriptions were used ($Z = -4.1$, $p < 0.001$), and in mentions of evoked context entities, where more {the} – properties descriptions than ϕ were used ($Z = -2.16$, $p < 0.05$). In mentions of evoked current entities more pronoun and ϕ expressions than {the} – properties realisations were used ($Z = -3.97$, $p < 0.001$). There was no significant difference in the syntactic forms used for mentions of displaced entities.

[5] A very regular feature of the end of long turns such as the ones described here is a diminution of pitch range and amplitude, yielding an overall "fade-away" effect. Over the last two or three intonation units of a turn we can indeed discern the near-monotone effect which Sanford and Garrod's model would predict for information which is in their terms "given" (cf. also discussion in Brown et al. [1980] of the end of long conversational turns). It is possible that length (duration) takes over from pitch/amplitude as the primary cue here.

It is as though once *triangle* is introduced into the discourse (in this type of discourse, at least, which demands a restricted vocabulary), the speaker resists marking the expression as requiring attention, even though the entity it introduces is new.

The second column in Table 6.1 shows the typical form of expression used to introduce inferable entities. These are very regularly introduced as definite noun phrases. These inferable entities (*middle, corner, side* of *page, apex, angle, side* of *triangle, radius, bottom, diameter* of *circle,* etc.) are presumably those which would be held to be introduced via the appropriate scenario [Sanford and Garrod, 1981, p. 112ff.] It is relevant to note that these expressions are typically associated with phonological prominence, i.e. in Halliday's terms, marked as new by the speaker rather than treated as given.

The third column, *evoked context* expressions, contains relatively few expressions (51 in all), most of which refer to *the page* the hearer is drawing on. Four refer to the pens provided, and the rest relate to the interlocutor *you.* All but one of the verbalised expressions are realised on low pitch without prominence. The one exception is from a speaker who begins by uttering the instruction "black pen", where both "black" and "pen" have associated pitch prominence.

I have divided Prince's last relevant category, *textually evoked,* into two subdivisions in order to show a quite marked difference in syntactic realisation between the subcategories. An item which has just been introduced into the discourse and which is currently the entity to which new information is being related, is here called a *currently evoked* entity. Items which have been introduced into the discourse at a point previous to the currently evoked item are here called *displaced*[6]. It is clear that the majority of forms relating to currently evoked entities are lexically "attenuated", to use Chafe's term (65% pronouns, 11% ϕ), whereas none of the forms referring to displaced evoked entities is lexically attenuated in this manner; indeed, a substantial proportion of expressions include not only nouns but adjectives identifying properties as well. It seems reasonable to suggest that, in general, a currently evoked entity might be regarded as readily accessible to a hearer, whereas a displaced entity may require specification in order to distinguish it from other, potentially competing, displaced entities. However, the level of lexical identification of displaced evoked entities in this data may result at least partly from the nature of the task – the need for specificity and the fact that as the task progresses there may be two entities of the same type which have already been introduced into the discourse and need to be distinguished. We might expect more pronominalising in expressions referring to displaced evoked entities in a narrative task, where different-gender participants are

[6] Yule [1981 b] develops a formal notation for tracking reference to entities through a discourse. I adapt his *current/displaced* distinction here without a properly developed discussion of the formal basis of his analysis.

involved, than in this task, where all the entities are same-gender. (A very similar incidence of lexicalised displaced entity expressions occurs in another genre where same-gender entities are mentioned, namely the description of routes from home to school/work [Brown et al., 1979].)

A striking feature of all the evoked entity-referring expressions in this data is that except for half-a-dozen examples of contrast under the evoked displaced column, all members of this general category are realised without pitch prominence. The pitch feature alone suffices to distinguish them from the typical realisations of the *brand-new* and *inferable* categories.

It is important to utter a caveat before making any general claims on the basis of this data. The texts produced by the speakers were mostly quite short – 150–200 words – and the topics they were concerned with were extremely limited. It is hardly surprising that the relatively few entities which are established in such brief texts should be expected by the speaker to remain highly accessible to the hearer, especially as the hearer has a visual record of those entities in the shape of the diagram he is in the process of drawing. It would clearly not be correct to generalise from this data and to claim that any entity referred to in any text will later be referred to by an expression which is uttered on low pitch. We have, for instance, conversational data which includes a text where A tells B about an elderly woman whom she met on the bus on her way to visit B. The talk drifts on to other topics. Some three minutes later, A reintroduces the elderly woman into the conversation with the expression "this lady" uttered with pitch prominence. It seems reasonable to suggest that the speaker judged that the "elderly woman" was no longer, in Chafe's terms, "present in the listener's consciousness" at the point when she wanted to refer to her again.

It is important, too, to note that the mere fact of repeating a referring expression does not constrain the speaker to uttering it on low pitch. Yule [1980] has provided a number of illustrations of a speaker who, having introduced a referent and repeated it low in pitch, may then immediately reinstate it high in his pitch range as he turns to further considerations of it:

even the dancing thing + dancing's no really a pastime
100 Hz 200–190 Hz

It is not the text sequence which determines how the speaker behaves, but the speaker's moment-to-moment assessment of the relationship between what he wants to say and his hearer's informational requirements. When he judges that the hearer is not expecting what he is about to say, he signals that the hearer needs to pay attention by using pitch prominence. When he judges that what he wants to say is well in line with the hearer's expectations, he drops the pitch down.

6.4 Conclusion

In the limited data described in the last section, the speaker's intonational behaviour accords well with Halliday's descriptions. When the speaker introduces brand-new information, he typically marks it with high pitch. When he introduces inferable information, he again typically introduces it with high pitch. The fact that inferable information is potentially available to his hearer through his activated schemata/scenarios, does not, most of the time, persuade the speaker to treat it as though the hearer expected it. It may be potentially "known" to his hearer, but the speaker treats it as new, in Halliday's terms. It is only the information which derives saliently from the context, or specifically from previous mention in the discourse, that the speaker treats as given, in Halliday's terms – on low pitch.

If we want to account for the speaker's use of intonation to mark the status of information for his hearer, Halliday's dichotomous categories "given" and "new" yield a satisfactory description of the intonational data. If, on the other hand, we wish to account for the range of syntactic forms associated with types of knowledge deriving from multifarious sources, clearly the categories of description require more than a simple dichotomy, and the terms "given" and "new" are stretched far beyond any possible utility. At least the range of distinctions drawn by Prince appears to be necessary, together with the current/displaced distinction which I have drawn upon here[7].

Acknowledgement. I am grateful to the participants at the Prosody Workshop and particularly to Henry Thompson for comments on this contribution.

---

[7] A recent paper by Marslen-Wilson et al. [1982] produces conclusions quite similar to mine within a different frame of reference; unfortunately it did not come to my attention until after the present contribution was completed.

7. Speakers' Conceptions of the Function of Prosody

Anne Cutler

With 2 Figures

7.1 Introduction

The place of prosody in language is, as the introduction to the present volume recounts, one of the most regularly disputed topics in linguistics. The question at issue is: must prosody be reckoned an integral aspect of linguistic structure, or is it in some sense overlaid upon an independent linguistic form? Many disputed linguistic questions, happily, can be enlightened by consideration of psycholinguistic evidence – evidence from patterns of language acquisition, from patterns of language breakdown, from production and comprehension performance – and this question would certainly appear to be one of those. In the present paper evidence from speech production is discussed, which, while not resolving the great issue of the place of prosody in language, does make a small contribution towards its resolution by allowing certain conclusions about what speakers themselves conceive to be the function of prosody in communication.

Fluent speech is accomplished in combination with fluent prosody; disfluencies of any kind in the speech stream can exercise effects on the prosodic contour. The evidence to be considered below concerns the interaction between prosody and one particular variety of speech disfluency, namely the commission of a speech error or slip of the tongue. In particular, the patterns of speech repair – the prosody of repairs, and the repair of prosody – are investigated, and, with respect to prosodic repair, the question of whether the occurrence of a repair is dependent upon the occurrence of an error.

The data on which the arguments are based are taken from the corpus of speech errors assembled by the author over the past eight years, and where appropriate, from other published speech error collections. The author's collection now includes a large number of tape-recorded errors, and the results presented in Sect. 2 derive from an analysis of this auditory corpus.

7.2 The Prosody of Repair

When a speaker perpetrates a slip of the tongue, either the error is noticed, or not. Only in the first case can we expect the presence of the error to affect the prosody. The speaker has two choices of how to react after detecting the error: either to correct it or not. If the speaker chooses not to issue a correction, it is nevertheless often apparent from momentary hesitation or from disruption of the rhythm of the utterance or of the pitch contour – i.e. from the prosody – that the speaker is aware of the fault. For instance, a speaker uttering

... like the Complex Noun PHRASE Constraint ...[1] (7.1)

failed to correct the erroneous assignment of stress to "phrase" instead of "noun", but betrayed his awareness of the error by faltering on the utterance of "constraint".

However, it is with the prosodic characteristics of error correction that the present section is concerned. Again, the speaker has two choices; that is to say, error corrections fall into two categories. In the first type of correction, the speaker (consciously or not) minimises the disruptive effect of the error on the utterance as a whole, and does not call the audience's attention to the remedial action. In the second type, the correction may be remarked upon explicitly, or in some other way assigned salience in the utterance. Goffman [1981] refers to these two types of correction as "flat" and "strident" respectively[2]; we will refer to them as "unmarked" and "marked". The main advantage of this more neutral choice of terminology is that we avoid the implication borne by the term "strident" that marking must always be achieved via *greater* amplitude and *higher* pitch; in fact, some corrections are clearly set off prosodically – "marked" – by being uttered on a noticeably lower pitch.

An unmarked correction is uttered as far as possible on the same pitch as the original error. This applies whether or not there is a pause preceding the correction, and irrespective of the amount of speech material repeated in the correction (but see below). For instance, both (7.2) and (7.3) were unmarked corrections:

... and bowls the first ball to Mike – Martin Kent (7.2)

That whole seem – (pause) – that whole bit seems .... (7.3)

In (7.2) the speaker, having produced the wrong first name, corrected with only a short pause, and on the same pitch as before. In (7.3), which is

[1] Syllables in upper case are ones which bear lexical stress or sentence accent, according to the context.

[2] Goffman distinguishes between the two types of correction on the level of speaker motivation: a speaker making a flat correction is not embarrassed at the slip, whereas a strident correction expresses the speaker's desire to convince the audience that the slip was neither intended nor characteristic, and to cancel it out as effectively as possible. Whatever the source of the difference, the two types of correction can be easily distinguished in any auditory speech error corpus.

probably an incomplete movement error in which the words "bit" and "seem" swapped places in the utterance, the speaker detected the error in midstream, paused, and then decided to backtrack to the beginning of the phrase. The first three words of the repeat phrase were uttered with the same pitch contour as the original erroneous three-word utterance. Figure 7.1 shows amplitude and pitch traces for (7.2), and it can be seen that error and correction are similar in both respects.

A marked correction, on the other hand, is distinguished by a quite different prosodic shape from that of the original utterance. Typically the correction material is uttered on a higher pitch and with greater intensity than the erroneous material. Examples (7.4) and (7.5) are marked corrections:

You're happy to – WELcome to include it. (7.4)

... then he himself loses the chance, that is, he RISKS the chance of dying.... (7.5)

Although the speaker of (7.4) seems to have detected an error more quickly than did the speaker of (7.5), and although the speaker of (7.5) included an explicit "editing term" [Levelt, 1983] – i.e. the phrase "that is" – while the speaker of (7.4) did not, in both cases the correction material – "welcome" in (7.4), "risks" in (7.5) – was stressed, as if to call the hearers' attention to the correct reading. Figure 7.2 shows amplitude and pitch traces for (7.5), and it can be clearly seen that the correction word is uttered both with higher pitch and with higher amplitude than the original production of the error word.

The markedness distinction, although apparently orthogonal to the time-course of error detection and correction, does interact with the *type* of error being corrected. Just as there is abundant evidence in the speech error literature that the pattern of occurrence of slips of the tongue is not random, so is there abundant evidence that the pattern of speech error correction is determined by a number of factors, touching both the nature of the slip itself and its effect on the discourse in which it occurs. Some types of errors are corrected consistently more often than others; some types are corrected in a consistently different way than others.

For instance, Nooteboom [1980] compared the correction patterns for phonetic errors (errors involving single phonetic segments) and lexical errors (errors involving whole words). He found that although both types of errors were corrected with approximately equal frequency, the form of correction differed across the two error categories in two ways:

(a) there was in general less backtracking – i.e. less repeated material – in the correction of a phonetic error than in the correction of a lexical error; and
(b) corrections of phonetic errors were in general faster – i.e. less material intervened between the first error segment and the first correction segment – than corrections of lexical errors.

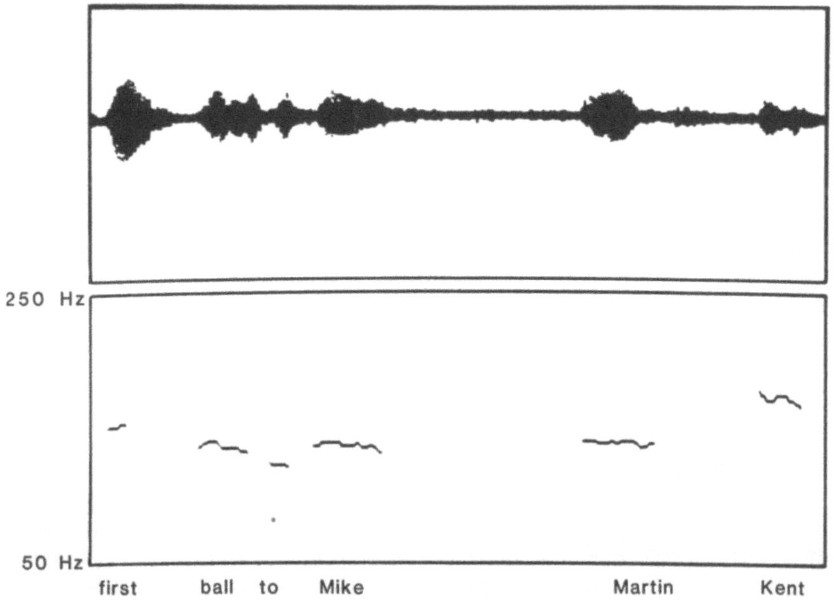

Fig. 7.1. Amplitude (*top trace*) and fundamental frequency (*bottom trace*) plotted against time for a typical unmarked correction. Error word and correction word are similar in amplitude, duration and pitch

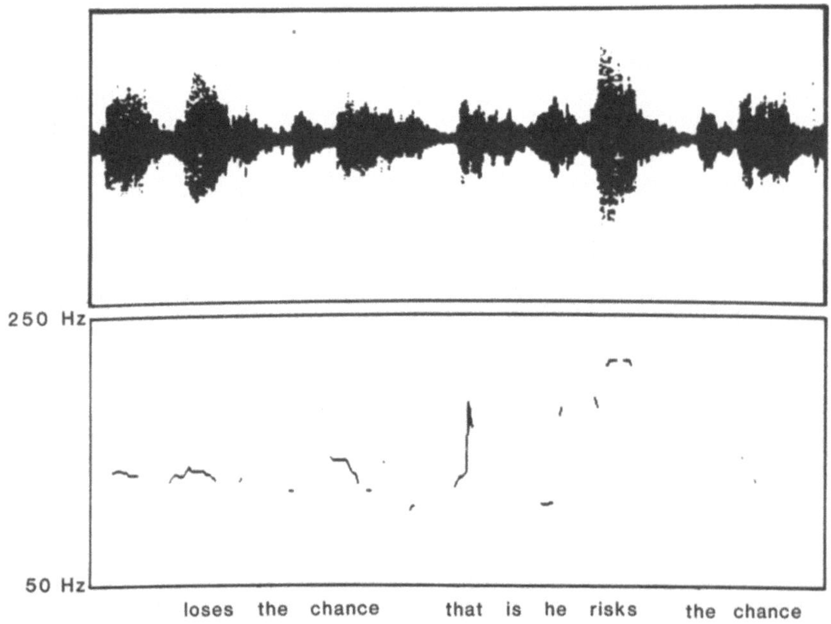

Fig. 7.2. Amplitude (*top trace*) and fundamental frequency (*bottom trace*) plotted against time for a typical marked correction. Correction word has greater amplitude, greater duration and higher pitch than error word

Nooteboom's interpretations of his data invoke the idea of an output-monitoring device as postulated by Laver [1973]. Laver's Monitor decodes utterances as they are spoken and checks them for

(a) linguistic well-formedness and
(b) success at communicating the intended message.

In fact, a monitor which attends to the spoken output only would have to be supplemented by a separate pre-output monitoring device, for which there is evidence in the finding that more real words than would be expected by chance result both from experimentally elicited slips of the tongue [Baars et al., 1975] and spontaneous slips [Dell and Reich, 1981]. Nevertheless the detection and correction rate for slips which have passed pre-output filters can be interpreted as a reflection of the operation of a monitor operating upon the speech output.

The lack of difference between phonetic and lexical errors with respect to frequency of correction is interpreted by Nooteboom as evidence that neither communicative success nor linguistic orthodoxy is an overriding criterion for correction. If the Monitor wanted above all to ensure that communication was not impaired, Nooteboom argues, lexical errors ought presumably to be corrected more often than phonetic errors. If, on the other hand, linguistic well-formedness were all-important, then phonetic errors, which often result in the utterance of non-words, ought to be corrected more often. Instead, claims Nooteboom, the Monitor apparently strives impartially for both adequate communication and conformity to linguistic form.

The later detection point for lexical errors (as reflected in the greater span of utterance before initiation of the correction) in comparison with phonetic errors is explained by Nooteboom in terms of the order of operation of the Monitor's checking routines: he suggests that the utterance is first checked for the linguistic orthodoxy of the phonetic form, and only later for syntactic and semantic appropriateness of units above the word level. The greater amount of repeated material in the correction of lexical as opposed to phonetic errors is assumed to reflect the relative size of the domain of phonetic versus lexical well-formedness. The domain of phonetic well-formedness is the word, so that corrections of phonetic errors only need to restart the word. The domain of lexical appropriateness, however, is at least the phrase (especially since choice of the wrong word may often produce fallout in accommodation of inflections on other words; see e.g. Garrett [1975]), so that backtracking to the beginning of the syntactic constituent is called for.

The same pattern of correction frequency, speed of correction, and amount of backtracking as reported by Nooteboom holds for phonetic and lexical errors in the auditory corpus being discussed here. However, the phonetic/lexical distinction is also reflected in an interesting way by the distribution of error corrections on the markedness dichotomy. Briefly, unmarked corrections can be of either kind, phonetic or lexical; but all of the marked corrections in the present corpus fall into the lexical category[3]. The difference is statistically significant.

Like the other differences which Nooteboom identified, the fact that lexical errors may be marked but phonetic errors are always unmarked would appear to be a function of domain – the domain, in this case, of accent placement. Marking a correction is, in effect, applying accent to it[4]. The function of accent, when it is applied to a lexical item which replaces another lexical item in an otherwise identical syntactic context, is to express contrast; thus the speaker of (7.4) could as well have said "not HAPpy, but WELcome", the speaker of (7.5) "not LOSes, RISKS". However, accent can only be applied to words – not to sounds. The replacement of erroneously produced sounds would thus not be expected to be marked, since sounds cannot be accented separately within a word. The function of accent being to highlight one word with respect to others in the same syntactic configuration, the domain of accent placement is necessarily the phrase. Marking, being the application of accent to a correction, is thus not an available option for the correction of phonetic errors, since the domain within which these are specified is too small to satisfy the conditions for the determination of accent placement.

This is nevertheless not the whole story. Although all phonetic error corrections are unmarked, many lexical error corrections are too – in the present corpus only 38% of lexical error corrections are marked. Inspection of the corpus reveals no obvious characteristic of lexical errors which correlates with marking; for instance, as mentioned above, the marked and unmarked corrections do not appear to differ on either of Nooteboom's measures (delay of correction or amount of backtracking). Nor is there a difference as a function of the type of error (omission, substitution, blend, etc.), nor as a function of the semantic relationship of error word to target word (substitution of a word from the same semantic field, as in (7.2), versus substitution of an antonym, for example).

It is possible that an extended sample of error corrections would reveal correlations not detectable in the present analysis. It is also possible – though

---

3 The lexical category here includes syllables (one marked correction) and morphemes (one marked correction). The phonetic category includes three prosodic errors (lexical stress and sentence accent). In all other errors single words or sounds were moved or replaced.

4 Marking may, as pointed out above, be realised in several different ways – by longer relative duration, noticeably higher or lower pitch, noticeably higher or lower amplitude, or a combination of pitch, amplitude and durational effects. Accent, similarly, is conceived to be an abstract entity which may have several alternative modes of acoustic realisation. (See the introduction to this volume for further discussion.) The terms "marked" and "unmarked" have not simply been replaced here by "accented" and "unaccented", however, since to do so would obscure the distinction between accent which derives from the normal process of prosodic structure assignment in speech production, and accent which is applied to mark a repaired production. Some corrections are accented but not marked – if a lexical error occurs on the word bearing the main accent in an utterance, then an unmarked repair will be similarly accented. A marked repair would assign *more* or *less* prominence to the correction word.

scarcely testable – that an explanation lies in the characteristics of the individual conversation in which each error occurred, i.e. that speakers mark a correction when they fear that the error is particularly likely to disrupt communication of their intended message. This, of course, would suggest a revision of Nooteboom's conclusions; ensuring unimpaired communication may indeed be somewhat more important to the Monitor than ensuring linguistic well-formedness. Interestingly, the present corpus includes several cases in which an initial unsuccessful unmarked attempt at correction was followed by a second, marked correction, as in (7.6):

The mechanism with which, er, at which – BY which an error occurs.... (7.6)

The next section, which deals with specifically prosodic errors, will return to this issue of successful communication, and also to the question of the domain over which different types of well-formedness are specified.

7.3 The Repair of Prosody

Certain speech errors consist simply of a distortion of the intended prosody – misapplication of lexical or compound stress, misplacement of phrase or sentence accent, inappropriate intonation contour, etc. [Fromkin, 1977; Cutler, 1980b]. These errors show an interesting pattern of detection and correction, namely: lexical stress errors are very often corrected, but the likelihood of correction is determined by the effect of the stress shift on the segmental structure of the word; prosodic repair of other than lexical stress errors, however, occurs only very rarely, and then only when the uttered prosody was not in itself incorrect, but merely inconsistent with the speaker's intention.

Lexical stress errors are corrected in approximately 50% of the cases in the author's corpus[5]. Examples (7.7–10) are typical lexical stress errors.

... from my PROsodic – proSODic colleagues . (7.7)

... each of these acoustic property detectors perhaps being subJECT – perhaps being SUBject to... (7.8)

You think it's sarCASm, but it's not . (7.9)

We're only at the early stages of it, we're still enTHUSiastic . (7.10)

[5] Correction means spontaneous correction by the speaker. In the other cases, the speaker's awareness of the correct stress pattern was established either by questioning, or by his production of the correct form at some other time.

When lexical stress mistakenly shifts, it falls on a syllable which is marked for stress in a morphological relative of the target word [Cutler, 1980 b]. Thus in (7.7) the first two syllables of "prosodic" were pronounced as they would be in "prosody"; (7.9) is reminiscent of "sarcastic", and so on. But the correction pattern for lexical stress errors seems to reflect only the speaker's assessment of whether or not the hearer could correctly identify the target word, i.e. whether communicative success was attained, because the likelihood of correction depends on the degree of distortion of the segmental structure of the target word. Specifically, in the present corpus stress shifts which result in a vowel quality change anywhere in the target word – full vowels replaced by reduced vowels or vice versa – are corrected in 62% of occurrences. Where no vowel quality change is involved, however, only 23% of shifts are corrected. The difference is statistically significant.

In fact, recognition of a spoken word is not always impaired by a stress shift when the word's canonical pattern of full and reduced vowels is maintained [Cutler and Clifton, 1983]. Speakers who commit lexical stress errors thus behave consistently with the perceptual facts: they correct such errors most often when word recognition is likely to be impaired. The output monitor, again, seems to consider efficient communication a more important goal than conformity with linguistic form.

The pattern of other prosodic repairs provides more subtle evidence of the same phenomenon. Examples (7.11–13) are cases where the speaker revised the initially chosen prosody.

... and what I'M saying – what I'm SAYing is ... (7.11)

... what recognisable circumstances – what REcognisable circumstances ... (7.12)

... despite the FACT – despite the fact that .... (7.13)

In (7.11), emphasis has been moved from one word to another; in (7.12), emphasis which was not applied in the first utterance has been added in the repetition; in (7.13), the converse has occurred and an initial emphatic utterance has been toned down. It is, however, immediately noticeable that the original prosody was not anomalous in English. The speaker has repaired not an incorrect output, but an output which was not what was wanted. It is impossible to determine whether the prosodic re-assignment resulted from the speaker's awareness of having committed an error relative to the original intention, or whether it resulted from a change of intention. Levelt [1983], in a discussion of all manners in which speakers may repair their speech, distinguishes between "error repairs" and "appropriateness repairs", the latter being changes of plan, in which the chosen wording is altered to an alternative which the speaker finds more appropriate. In Levelt's terms, then, cases like (7.11–13) are ambiguous – they could be either error repairs or appropriateness repairs.

Whatever the nature of such repairs, what appears to precipitate them is the possibility that hearers might construe the utterance in a way other than what the speaker wants. Sentence accent is not fixed in English; it can fall on practically any word in many utterances, and changes in the location of sentence accent correspond to changes in the pragmatic force of the utterance. Thus an acceptable but unintended accent pattern could result in the audience misinterpreting the implications of the speaker's utterance, and to avoid such misinterpretation, the speaker revises the prosody.

Anomalous accent placement seems, however, not to bother speakers at all, as the examples in (7.14–20) demonstrate.

We're also going to be talking about Citizens' BAND Radio. (7.14)

... especially in the universe within WHICH we're looking. (7.15)

She had a lot of cups, but the one SHE gave me leaked. (7.16)

The only trouble WITH it – (pause) – is the hood is too small. (7.17)

The subject of an embedded clause can be connected to the subject of a higher CLAUSE. (7.18)

... the effects of visual presentation withOUT distraction and auditory presentation without disTRACtion. (7.19)

... from three hundred THOUsand, which used to be the norm, it's now one point FOUR million. (7.20)

These utterances incorporate a variety of anomalies. (7.14), for example, involves incorrect phrase accent – compare (7.1), in which incorrect compound stress is assigned within a phrase. In (7.18) the speaker has failed to deaccent a repeated lexical item. In (7.19) contrastive accent has been incorrectly placed – the speaker should presumably have contrasted "visual" and "auditory". In (7.16) accent falls upon a pronoun, as intended – but on the wrong pronoun. The utterances have in common only that the accent pattern is in each case anomalous, and that none of them was corrected.

This non-correction pattern could arise in two ways: either errors like (7.14–20) are simply not detected by the output monitor, or they are detected but they fail to satisfy the monitor's criteria for correction. There is some evidence that each of these explanations holds for certain errors.

The question of detection of such errors raises once again the notion of the domain of well-formedness of linguistic phenomena. It is noticeable that many uncorrected accent placement errors are ill-formed only with respect to quite large units of discourse – more than the phrase in which they occur, often more than the clause or sentence. Thus each of the phrases in (7.19) would be appropriately accented in a different context; only in conjunction is the anomaly apparent. The same is true of (7.18) and (7.20) (and of dozens of similar uncorrected deaccenting and contrast errors in the author's collection). Furthermore, uncorrected focus errors, unlike corrected ones such as (7.11–13), are errors only with respect to the extra-sentential context:

So THIS sort of thing happens in other fields. (7.21)

This was the latest in a series of one-day strikes; another will be held on NEXT Tuesday. (7.22)

It is only clear that the accent placement in (7.21) is anomalous when one knows that the speaker, having earlier described a fault typical of psychologists, used these words to conclude a digression which had exemplified the same fault in the field of chemistry; accent should have been applied to "other". Similarly, (7.22) is only odd if one knows that the strike being described had been held the same day that this sentence was uttered, and that that day happened to be a Friday. Neither (7.21) nor (7.22) was corrected.

Thus it indeed appears possible that the explanation for the non-correction of many accent placement errors is simply that they are not detected because the stretch of discourse with respect to which they are anomalous is larger than the domain available to the monitor. This explanation can be tested by checking the correction pattern of other types of errors in which ill-formedness depends on a non-local context. In fact, a large class of such errors exists; Brown [1980] has termed them "grammatical incoherence". (7.23–28) are examples of such errors from my own collection.

Work on identifying features of the autonomous plane are not that far advanced (7.23)

People like Posner with whom one associates sophisticated theories of attention is doing some work (7.24)

Remember that wide-angle lens that I was going to borrow it? (7.25)

That was something on which the theory, up to how we'd formulated it so far, was silent about (7.26)

. . . because I have a filter that throws out everything around a thousand Hertz out (7.27)

. . . nor did we say that it was the standard to which instructors should teach to. (7.28)

Many such errors cited by Brown, as well as many in my own collection, involve accommodation of verb inflections to the wrong antecedent – e.g. (7.23) and (7.24) – or double occurrence of verb particles, as in (7.26–28)[6]; it

[6] See Fay [1980] for a discussion of the sources of such errors.

is worth pointing out that neither verb inflection errors nor verb particle errors are generally left uncorrected, as (7.29–32) show:

...what things are this kid – is this kid going to say incorrectly (7.29)

Where do these upside-down F's comes from – come from? (7.30)

Well, let me write it back, er, down, so that... (7.31)

I can't work out where I ran over – ran across that other name. (7.32)

In fact, (7.29) is essentially the same error as (7.23), except that the item governing the inflection occurs adjacent to the verb in (7.29), but eight words before it in (7.23). This point is neatly reinforced by (7.33), in which the speaker makes two inflectional errors in quick succession; the first, adjacent to its governor, is corrected, but the second, governed by a word much earlier in the sentence, is not:

The place where their rules does – do seem to come in handy are when you have.... (7.33)

Thus the failure to correct (7.23–28) seems to be associated primarily with the size of the linguistic unit within which the error is defined[7], and provides independent confirmation that the failure to correct accent placement errors such as (7.14–20) results from the same cause. Precise specification of the domain over which ill-formedness can be detected is probably best achieved by explicit experimental investigation of this question. However, some indirect evidence is available from the previous studies of error correction cited above. Nooteboom [1980] reports that in his corpus the chance of an error being detected seems to have dropped to zero by about five words after the error has occurred. Levelt [1983] describes a slightly more complicated picture – correction of an error which occurs at a syntactic boundary is more likely than correction of an error which occurred in mid-phrase. Together these observations suggest that the monitor which is responsible for the detection of speech output in need of repair operates, roughly, phrase by phrase. Since the monitor is essentially a comprehension device – and Levelt [1983] argues specifically that monitoring is carried out by the same parsing device which we use to understand the speech of others – it is tempting to compare its limited domain with the limits postulated for models of comprehension. Frazier and Fodor [1978], for example, argue for a "window" length of six words for their parser. The error correction evidence itself, however, indicates only that the monitor has a very limited, and most probably syntactically defined, domain of operation; and that its domain is prob-

[7] My collection of such errors is much smaller and has been much less systematically assembled than my collection of prosodic errors, so that a statistical test of this suggestion is inappropriate on the present corpus.

ably smaller than the domain available to the listener's comprehension device (anomalous accent placement is, after all, often obvious to the hearer).

The non-correction of many accent placement errors, then, may most likely be attributable to the fact that they are simply not detected because the domain of their ill-formedness is larger than the domain of operation of the output monitor. It has already been noted, however, that accent placement often *is* repaired – as, for example, in (7.11–13). It is noteworthy that these repairs are issued comparatively rapidly – in my collection the maximum number of words intervening between the word which is to be accentually repaired and the beginning of the repair is (in one case) three, and the mode occurs at one word. This pattern suggests that focal accent may be defined over a comparatively short domain, short enough for the monitor to be able to determine that focus was not placed where intended, or could have been more appropriately placed. Of course, it is possible to speculate that it is precisely the cases in which focus is applied within a short constituent which are detected and repaired, and that there may be many more such cases in which speakers' original intentions for focal accent placement are not realised in their utterances, but the misplacements are detected neither by the speakers' own monitors nor – since sentence accent is not fixed – by the audience. Perhaps this question may in due course also prove susceptible to experimental study.

Finally, however, there are still some cases in which accent misplacement is apparently detected, but no repair is issued. The speaker's pause in (7.17), for instance, seems to indicate consciousness of a problem, just as the post-error faltering in (7.1) indicated that that speaker had noticed his faulty compound stress placement. Since the domain of compound stress assignment is the nominal compound itself, i.e. two words, there is every reason to expect (7.1) to have been detected. Similarly, (7.17), in which accent should have been applied one word earlier, should have been as easy to repair as (7.11–13). The speakers' failure to repair in these cases seems, in terms of our earlier discussion of the monitor's priorities, to imply that communicative success did not seem to be threatened. In other words, prosodic repairs are issued only when – as in (7.11–13) – the speaker wishes to correct the message being provided to the hearer. Anomalous compound stress placement, as in (7.1), and assignment of accent to a function word, as in (7.17), do not suggest a particular interpretation of the utterance which is different from the interpretations obtaining had accent fallen upon "noun" and "trouble" respectively. Therefore, one may hypothesise, the speaker did not feel it necessary to correct the error.

Of course, it is often the case that anomalous contrast placement and failure to deaccent will likewise have no effect on the pragmatic implications of the utterance; although the evidence cited above indicates that most such errors cannot be detected, it is still possible that some are detected but that the speaker chooses not to issue a correction, relying instead on the hearer's ability to construct an appropriate prosody from the context.

7.4 Conclusion

The interactions between prosody and speech repair illustrate speakers' conceptions of the functions of prosody. Firstly, speakers use prosodic means to ensure that errors which are likely to disrupt communication are decisively repaired. An error correction which is "marked", i.e. is distinguished prosodically from the original utterance of the error, is only applied to an error at the lexical level or above. Marking ensures that the hearer attends to the correction; the function of prosody in this case is to direct the hearer's attention.

Secondly, prosodic repairs are issued when the speaker fears the hearer might be misled into an inappropriate interpretation of the utterance. Anomalous accent placement itself, as long as it does not carry unwanted pragmatic implications, is not corrected. The function of prosody is thus seen to be, in the speaker's view, primarily concerned with the semantics or pragmatics of the utterance. Similarly, anomalous lexical stress placement is rarely corrected unless it is accompanied by alteration of the segmental structure of the underlying word. Lexical stress thus seems to be perceived by speakers as inessential for communication; maintenance of the word's canonical pattern of full and reduced vowels is what is important for word recognition.

In this speakers are behaving in accord with the perceptual facts, since, as pointed out above, perceptual studies show that incorrect stress placement, while often obvious to the hearer, does not always disrupt word identification unless full vowels are replaced by reduced vowels or vice versa. And indeed, there is evidence that speakers' conceptions of the function of accent are equally realistic. Studies of the perception of accent show that listeners actively search for the accented words in an utterance [Cutler, 1976], and that this reflects their search for the semantically most central portions of the speaker's message [Cutler and Fodor, 1979]. Speakers' conceptions of the functions of prosody, therefore, seem to be in considerable accord with psycholinguistic reality.

Acknowledgement. This research was supported by a grant from the Science and Engineering Research Council to the University of Sussex. I am grateful to Dwight Bolinger, Jack Hawkins, Steve Isard, Bob Ladd, Ewald Lang and Pim Levelt for helpful comments.

8. Structures and Categories in Prosodic Representations

Daniel Hirst

With 5 Figures

8.1 Intonation and Universal Grammar

Intonation, what Bolinger has called the “greasy part of language”, is notoriously difficult to describe. One of the reasons for this elusiveness is of course the essentially spoken character of intonation; a deeper reason, however, comes from the fact that an adequate description of intonation needs to take into account not simply the phonology of the language, but also the syntax and the semantics, as well as the interfaces between the grammar and the “real world” constituted by phonetics and pragmatics.

One of the fundamental questions that linguists have long asked concerns the way in which intonation contributes to the “meaning” (in the widest sense) of an utterance[1]. That no satisfactory simple answer to this question has so far been forthcoming is in itself an indication of the complexity of this issue.

Within the framework of Universal Grammar as described, for example, in Chomsky [1981], it is assumed that a grammar is organised along the following lines:

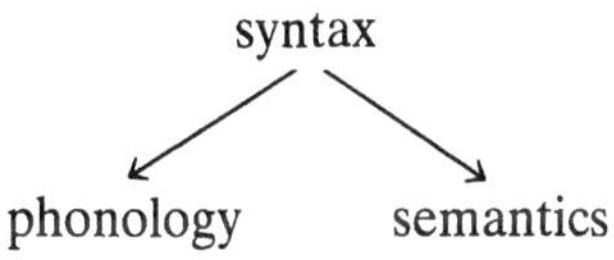

(8.1)

Given such a framework, a theory of intonation will need to include at least the following subcomponents:

a phonological component specifying the structures and categories of underlying representations (8.2 a)

---

[1] For an attempt to answer this question by means of an enriched deep structure, see Hirst [1977]. It will be apparent from this paper that I no longer accept the basic thesis of my earlier study (i.e. that intonative meaning should be encoded in deep structure), although many of the descriptions presented there can be translated directly into the sort of interpretative framework I assume here, and I still hold that a crucial distinction is to be made between the “functions” of intonation and the forms that embody these functions.

a phonetic component defining the relationship between the underlying representation and the acoustic parameters (i.e. segmental duration, fundamental frequency and intensity as functions of time) (8.2 b)

an interpretative component relating the underlying representation to syntactic, semantic and pragmatic aspects of meaning. (8.2 c)

It will be evident from the above that the problem of *representation* is crucial to the study of intonation. In this contribution I shall essentially be concerned with the nature of phonological representations of prosody. I shall, however, try to show that the nature of the phonetic component (8.2 b) has important consequences for the way in which we formulate (8.2 a) and, by way of consequence, (8.2 c). After a preliminary description of the phonetic component in Sect. 8.2, I make some proposals concerning prosodic representations in Sect. 8.3, and in the light of these proposals, I return in Sect. 8.4 to a brief discussion of remaining problems of sound and meaning.

8.2 Phonetic Representations of Intonation

Variations in pitch are essentially determined by the fundamental frequency (F_0) measured in hertz (cycles per second)[2]. A very low level representation for pitch can consequently be provided by a sequence of F_0 values at a given sampling rate. For speech synthesis, for example, a common sampling rate is 100 values per second. Such a representation, however, obviously contains a great deal of redundant information. In order to eliminate the redundancy as a first step towards a phonetic rather than an acoustic representation, we need to develop a model for pitch changes.

[2] As mentioned in Sect. 8.1, a complete model will need to specify segmental duration and intensity as well as fundamental frequency. There is some disagreement as to the relative importance of these acoustic parameters, but in an experiment described in Faure et al. [1980; written in 1974!] we found that with minimal pairs for word stress (such as "green house" versus "greenhouse"), while listeners achieved near perfect scores when the words were presented in the context "It's a ___", they achieved no better than random scores when the words were presented in the context "I know very *well* it's a ___" with the major pitch change on "well" and low static pitch throughout the rest of the sentence. Further investigation, however, revealed that the total duration of the Adjective + Noun ("green house") expressions was on average over 20% longer than that of the compound word ("greenhouse") and that this difference in duration was statistically significant ($t = 5.196$, $p < 0.001$). Thus the speaker produced systematic differences in duration for the two types of expressions although these differences were apparently not sufficient for listeners to detect. This could be taken as evidence that while durational differences are a primary correlate of stress differences at the level of speech production, pitch changes are predominant at the perceptual level.

One of the first problems for such a model is the fact that fundamental frequency is not a continuously observable parameter of the sound wave, its presence or absence being dependent on the segmental feature of voicing. This means that the following sentences, for example, in French:

C'est papa? /sepapa/ "Is it daddy?" (8.3 a)

C'est maman? /semamã/ "Is it mummy?" (8.3 b)

would not have the same *acoustic* representation (Fig. 8.1), whereas we should presumably expect them to have the same *phonetic* representation,

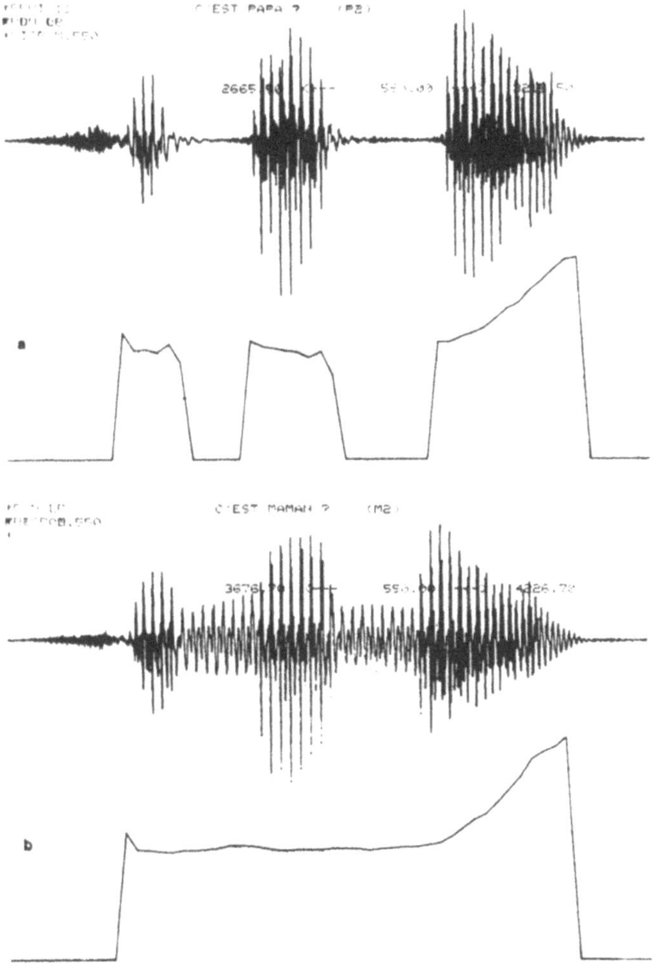

Fig. 8.1 a, b. Waveform (*upper line*) and F_0 (*lower line*) for the sentences (**a**) "*C'est papa*?" and (**b**) "*C'est maman*?"

and a fortiori the same *phonological* representation as far as their intonation is concerned.

The difference between the F_0 curves in Parts (a) and (b) of Fig. 8.1 can be described in two rather different ways, depending on which of the two curves we take to be the unmarked case. On the one hand, we could adopt a segmental approach and model the F_0 curve for each vowel, assuming that the fundamental frequency on voiced consonants is simply the result of interpolating between the adjacent vowels. On the other hand we could adopt a suprasegmental approach, taking F_0 to be an underlying *continuous* function of time, and assume the observed discontinuity to be the result of the interaction between this underlying function and the voiced/unvoiced nature of the successive segments. Either model would adequately describe the actual curves of the sentences in Fig. 8.1 a, b; however, this is not always, or even usually, the case. Figures 8.2 a and 8.3 a represent the sentence:

Il n'a pas vu le chien fou. "He didn't see the mad dog." (8.4)

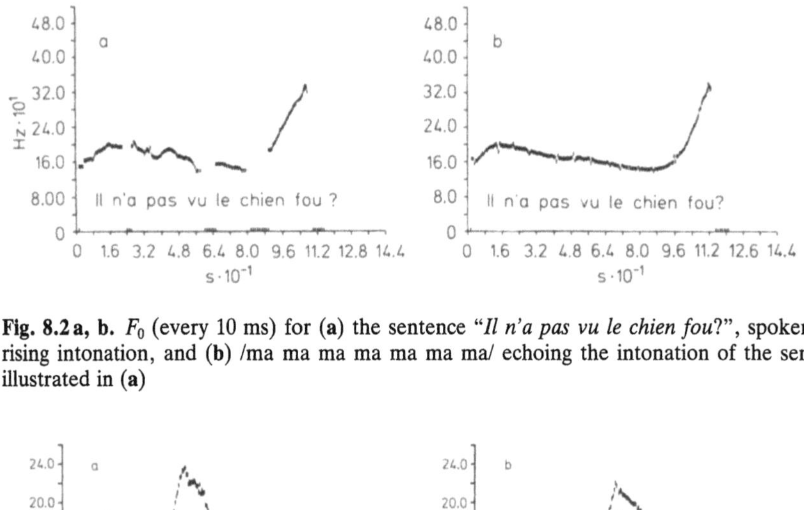

Fig. 8.2 a, b. F_0 (every 10 ms) for (**a**) the sentence "*Il n'a pas vu le chien fou*?", spoken with rising intonation, and (**b**) /ma ma ma ma ma ma ma/ echoing the intonation of the sentence illustrated in (**a**)

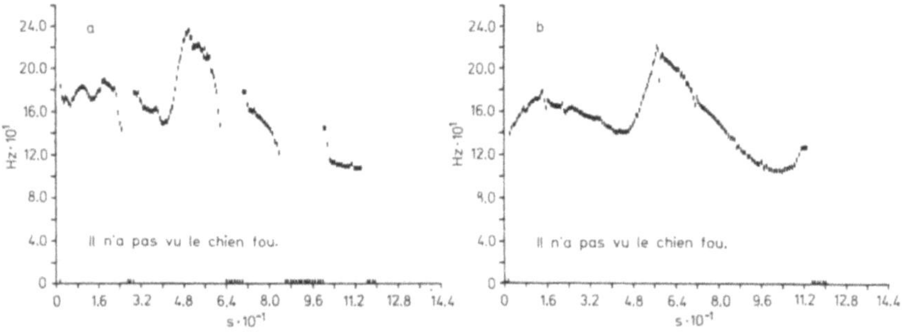

Fig. 8.3 a, b. F_0 (every 10 ms) for (**a**) the sentence "*Il n'a pas vu le chien fou*", spoken with falling intonation, and (**b**) /ma ma ma ma ma ma ma/ echoing the intonation of the sentence illustrated in (**a**)

with rising and falling intonation respectively. Figures 8.2 b and 8.3 b represent the same intonation as the sentences in Figs. 8.2 a and 8.3 a, echoed on a meaningless sequence of syllables "ma ma ma ma ma ma ma"[3]. A comparison of the a and b sentences shows that the model representing F_0 as a continuous function of time is clearly superior to the discontinuous model, since in Figs. 8.2 a and 8.3 a, after each non-voiced segment, the F_0 curve takes up again at precisely the point which the curve *would* have reached in the case of a continuous function (as can be seen by comparison with the b sentences), making allowance for some small local microprosodic variations.

I have suggested elsewhere [Hirst, 1981 a, c] that an F_0 curve can be analysed as the result of the interaction between voicing and the continuously evolving tension of the vocal cords. I also suggested that continuous F_0 curves can conveniently be modelled as a continuous, smooth sequence of parabolic arcs: technically, a quadratic spline function. The phonological interest of such a function is that each section of it is locally determined independently of the other sections. The curve is determined in its entirety by a sequence of *target pitches* defined on a partitioning of the curve[4].

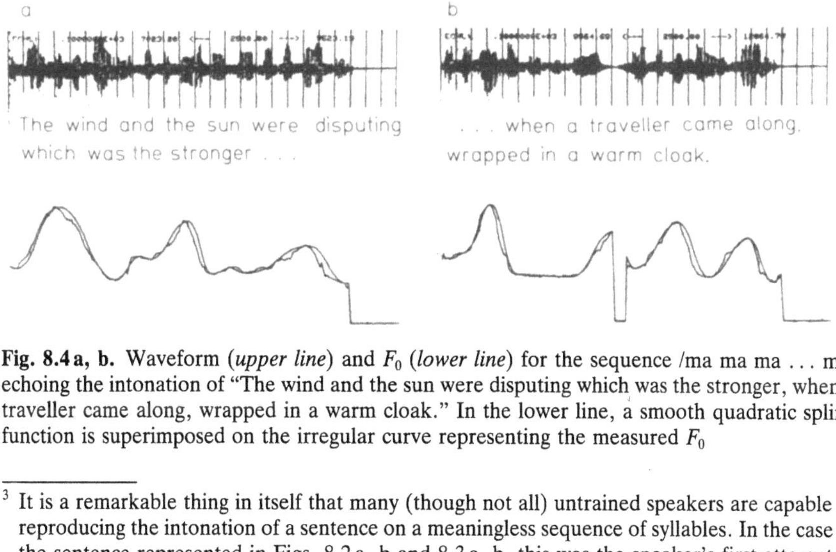

Fig. 8.4 a, b. Waveform (*upper line*) and F_0 (*lower line*) for the sequence /ma ma ma ... ma/ echoing the intonation of "The wind and the sun were disputing which was the stronger, when a traveller came along, wrapped in a warm cloak." In the lower line, a smooth quadratic spline function is superimposed on the irregular curve representing the measured F_0

---

[3] It is a remarkable thing in itself that many (though not all) untrained speakers are capable of reproducing the intonation of a sentence on a meaningless sequence of syllables. In the case of the sentence represented in Figs. 8.2 a, b and 8.3 a, b, this was the speaker's first attempt at reiterating intonation patterns, a feat which he had believed himself incapable of performing. For a recent discussion of the relevance and reliability of this well-known technique for prosodic studies, see Liberman and Streeter [1978].

[4] A spline function of degree n is defined as a continuous sequence of polynomial arcs of degree n, the derivatives of which, up to degree n − 1, are everywhere continuous. For an introduction, see Burden et al. [1981, Chap. 3]. For a quadratic function, each arc of the curve requires three parameters ($y = a_0 + a_1x + a_2x^2$) which can be determined from the target point or vortex of the parabola, and the time value of the juncture point with the following arc. For a more detailed presentation and a discussion of the appropriateness of this and various other possible models, see Hirst [1981 a, in press].

Figure 8.4 gives an example of the sort of fit obtained by such a function and the intonation of:

The wind and the sun were disputing which was the stronger,
when a traveller came along, wrapped in a warm cloak. (8.5)

reiterated on a sequence of "ma ma ma..." syllables. Using values obtained from such an analysis for speech synthesis gives results which are practically indistinguishable from the sentence resynthesised with the natural (measured) F_0 curve.

Given such a model, then, an appropriate *phonetic* representation for an F_0 curve would be a sequence of target pitches $\langle t_i, h_i \rangle$, where t_i represents the time value and h_i the fundamental frequency for each target pitch. Pierrehumbert [1980], in her recent analysis of English intonation, has claimed that there is *no* well-defined level of representation between the actual F_0-curve and the phonological tone sequence. Her argument, however, depends on a specific analysis of *downstepped tones* which leads her to posit two distinct mechanisms of pitch interpolation. I shall try to show (Sect. 8.3.3) that a different analysis of downstep [Hirst, 1979, 1981 b] leads to a more unified treatment of pitch interpolation on the basis of a sequence of target pitches[5].

8.3 Phonological Representations

Given the conventions for a phonetic representation of intonation proposed above, the intonation of a sentence like "ma ma ma ma", reiterating the intonation of "a cup of tea", can be represented, for example, as follows:

$$\langle\langle 0.04\,\mathrm{s},\ 105\,\mathrm{Hz}\rangle,\ \langle 0.31\,\mathrm{s},\ 175\,\mathrm{Hz}\rangle,\ \langle 0.46\,\mathrm{s},\ 156\,\mathrm{Hz}\rangle,\ \langle 0.67\,\mathrm{s},\ 85\,\mathrm{Hz}\rangle,\ \langle 0.73\,\mathrm{s},\ 80\,\mathrm{Hz}\rangle\rangle \quad (8.6)$$

Figure 8.5 shows the segmental and prosodic timing for an actual realisation of this.

The phonetic component of the grammar will need to derive something like (8.6) from a phonological representation of the sentence. This derivation will involve two presumably distinct operations:

(a) the determination of the F_0 of the target pitches from the underlying prosodic features
(b) the determination of the timing of the target pitches with respect to the rest of the phonetic material

[5] Ladd [in press and this volume] comes to a similar conclusion concerning Pierrehumbert's representation of downstepped tones.

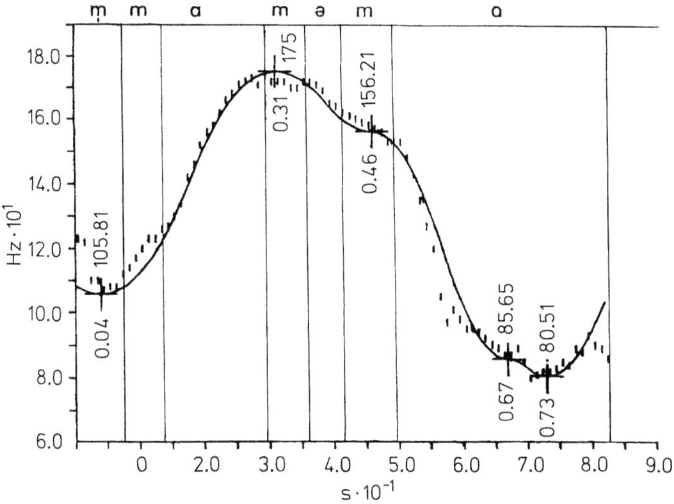

Fig. 8.5. Segmental and prosodic timing for the sequence /m ma mə ma/ echoing the intonation of "a cup of tea." The small vertical lines represent the measured F_0 (every 10 ms), while the continuous line shows the fit obtained by a quadratic spline function through the target points ⟨0.04 s, 105.61 Hz⟩, etc.

8.3.1 Phonological Tones

I shall assume that the pitch values in (8.6) are derived from a sequence of *tones* defined by a restricted set of phonological pitch features. I have argued [Hirst, 1979] that the same set of features used to describe the pitch characteristics of tone languages can satisfactorily be applied to a language like English. The stepping intonation typically described for unmarked declaratives in British English [O'Connor and Arnold, 1961] is, as has often been noted, quite similar to the widespread phenomenon of *downstep* observed in a large number of African tone languages such as Akan and Bambara. Thus Schachter [cited in Stewart, 1971, p. 196], has remarked that

> the Twi [the Akuapem Twi dialect of Akan] tone system can produce counterparts of virtually all the English intonation patterns.... For example the pattern of *He came this morning* is analogous to that of a Twi sentence
> ɔ̀'bɔ́'fónó 'húùnɔ̀ "The messenger saw him" .

Assuming just two pitch features HI and LO, I proposed to account for the downstepped tones by the combination + HI, + LO. I shall assume this analysis below, although it is not crucial to the rest of the argument. Taking the four possible combinations of features as follows:

Tones	Features
H(igh)	+HI, −LO
L(ow)	−HI, +LO
D(own)	+HI, +LO
M(id)	−HI, −LO

(8.7)

it would be possible to represent the sequence underlying (8.6) as:

$$\langle M, H, D, L, L \rangle \tag{8.8}$$

I shall not attempt to account for the actual derivation of pitch values from the tonal sequence, although it is perhaps worth mentioning that if we assume that pitch values are defined relative to a steadily declining baseline[6] within each utterance, there seem to be reasonable grounds for supposing that speakers in fact make use of quantal differences in pitch corresponding to binary divisions of the octave [Liberman, 1975; Hirst, 1981 c].

Taking (8.8) as the underlying sequence of tones for (8.6), then, I now examine the relationship of the tones to the phonological structure before returning in Sect. 8.3.3 to the question of the underlying form of (8.8).

8.3.2 Phonological Structure

The phonetic representation in (8.6), like the sequence of tones in (8.8), is clearly incompatible with the linear models of phonology developed in early versions of generative grammar, where a phonological marker is represented either as a string of phonemes [Chomsky, 1955, p. 66] or as a "two-dimensional matrix in which the rows stand for particular features; the columns stand for consecutive segments" [Chomsky and Halle, 1968, p. 5]. Such a model was, of course, an extremely strong hypothesis limiting phonological representations to a concatenated string of elements drawn from a single category, constituted by a simple unstructured data type: the phoneme or segment.

Such a segmental account will obviously run into difficulties in representing the intonation of a sentence like that illustrated in Fig. 8.5, since the first three target pitches are located on consonantal segments /m/, three segments seem to have no pitch value defined for them, and the final segment /a/ apparently has two consecutive pitch values. The particular values of the target pitches shown in Fig. 8.5 are of course open to dispute, but the example is surely in no way exceptional. Had we taken an example of a rising or

[6] See Vaissière [this volume] and studies cited there for an account of declination in various languages.

falling-rising intonation pattern, the incompatibility with a segmental analysis would have been all the more apparent.

There has been a growing awareness in the last decade that the hypothesis embodied in the linear approach to phonology was too strong, and there have been various proposals in the literature for extending the phonological model. One of the most common has been to enrich phonological representation by the introduction of a *structured* phonological category: the syllable [Wang, 1967; Venneman, 1972, 1978, 1979; Hooper, 1972; Kahn, 1976]. Note, however, that a syllable-based phonology does not solve the problems we saw above. If we return to Fig. 8.5, we see that while the first and third syllables are each associated with a single pitch value, the second syllable is associated with none, and the fourth (final) syllable is associated with three values. Another surprising difficulty for a syllabic account is the fact that while it is generally assumed that the pitch values are in some way connected with the stress pattern of the utterance, in Fig. 8.5 the first two pitch values occur precisely on the *un*stressed syllables, while the pitch on the first stressed syllable is simply an interpolation between the neighbouring values.

A more promising representation would seem to be obtained by introducing a further category, the *stress foot* [Selkirk, 1978 a, 1980], although there is still no one-to-one correspondence between the phonological categories and the tonal features.

The last few years have seen a considerable spate of interest in various *non-linear* models of phonological representation (for a useful overview see McCarthy [1982]). While it is always difficult to generalise about such a rapidly evolving field, the following principles seem to have attained some degree of consensus:

> Phonological representations should contain separate *lines* for tonal and segmental features, the two lines being associated by some form of phonological structure [Goldsmith, 1976; Halle and Vergnaud, 1982]. (8.9 a)
>
> Phonological structures consist of binary branching trees in which each branch consists of one *dominant* (strong) and one *recessive* (weak) node [Liberman, 1975; Liberman and Prince, 1977; Vergnaud and Halle, 1978]. (8.9 b)
>
> Within a given phonological category, all structure is either uniformly right-branching or uniformly left-branching [Stowell, 1979]. (8.9 c)
>
> Recessive nodes do not branch [Hayes, 1981]. (8.9 d)

I have argued [Hirst, 1981 b] that trees constructed according to (8.9) are in fact isomorphic with multiply branching trees in which each node is given a category label and each category is specified for either final or initial prominence. While Liberman and Prince [1977] did not explicitly introduce phonological categories into their representations, their use of prominence markings interacting with left/right branching can be interpreted as a means of implicitly labelling certain nodes as *Foot.* Once such a category is explicitly introduced, the prominence markings and left/right branching[7] become redundant in practically all cases, being predictable from the general principles in (8.9), and the following:

Within a foot the first syllable is more prominent than the others. (8.10 a)

Within a word or phrase the final foot is more prominent than the others. (8.10 b)

The only apparent exception to (8.10 b) in English occurs when a word ends with a final monosyllabic foot. Thus while in

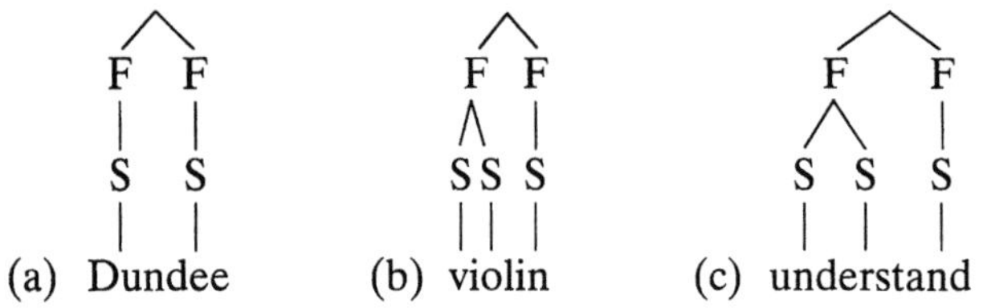

(8.11)

the final foot is dominant in accordance with (8.10 b), in

(8.12)

it is the penultimate foot which has the greatest prominence. It is obvious, however, that it is the words in (8.11) rather than those in (8.12) which are the exceptions. There appears to be a general principle (cf. Schane's "detail rule" [1972], Vergnaud and Halle's "super- (or "hyper-) syllable" [1978], Dogil's "trochaic projection constraint" [1981] and Hayes's principle of "extrametricality" [1979, 1981]) which prevents a final monosyllabic foot from being dominant in English. I suggested [Hirst, 1981 b] that we interpret this constraint as a general principle of word formation in English, allowing an extra foot to be formed over the last two feet of a word just when the final foot is monosyllabic. Thus the examples in (8.12) will be represented:

[7] This does not, of course, exclude the possibility that empirical evidence for such a distinction might be forthcoming.

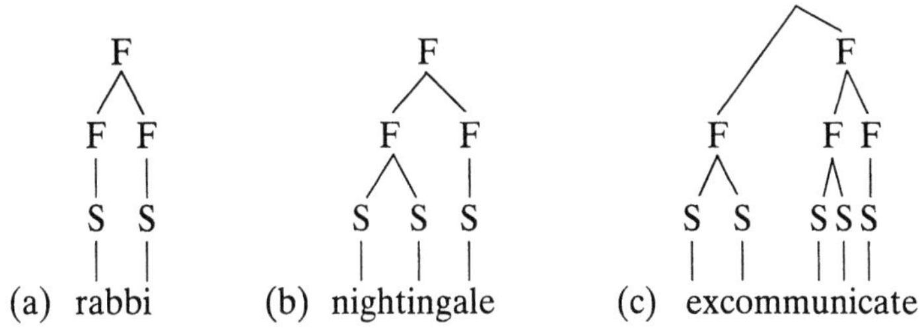

(8.13)

The principles in (8.10) will now be sufficient to determine correctly the relative prominence of the syllables in the representations in (8.11) and (8.13).

In the light of the above, I shall assume in the rest of this contribution a very simple model of phonological structure, conceived of as a multiply branching tree with each node marked P(hrase), F(oot) or S(yllable) and with tonal and non-tonal segments as terminal elements. The sentence illustrated in Fig. 8.5, for example, will consequently be represented by the segmental line (8.14a), and the tonal line (8.8) repeated here as (8.14b):

$$\langle \text{m}, \text{m}, \text{ɑ}, \text{m}, \text{ə}, \text{m}, \text{ɑ} \rangle \tag{8.14a}$$

$$\langle \text{M}, \text{H}, \text{D}, \text{L}, \text{L} \rangle \tag{8.14b}$$

where in (8.14a) "m" is short for [−voc, +cons, +nas...], etc. and in (8.14b) "M" is short for [−Hi, −Lo], etc. The two lines, let us assume, are associated by the structure:

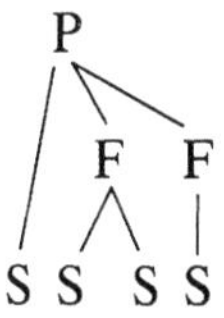

(8.15)

It is worth emphasising that under this interpretation, the tonal and the non-tonal segments are not, in fact directly connected at all. This is rather different from Goldsmith's [1976] "autosegmental theory", for example, where vowels and tones are directly associated according to

Well-formedness condition [Goldsmith, 1976]

i) All vowels are associated with at least one tone. All tones are associated with at least one vowel. (8.16)

ii) Association lines do not cross.

Within the framework I am proposing, the tonal (and non-tonal) segments are associated with a sequence of categories, corresponding to an exhaustive *segmentation* of the phrase [Clements, 1976, p. 5]. Furthermore, the introduction of *structured* phonological categories removes much of the motivation for representations of the form

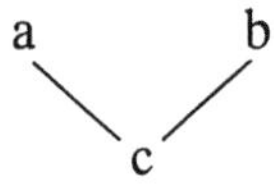

(8.17)

where a and b are phonological categories and c is a (tonal or non-tonal) segment. Suppose, then, that we explicitly exclude representations such as (8.17); the well-formedness condition will now be:

> Well-formedness condition (revised)
>
> At each level of representation there exists an exhaustive segmentation of the phonological structure into a sequence of categories such that:
>
> i) each category is associated with at least one segment (8.18)
> each segment is associated with *exactly* one category
>
> ii) association lines do not cross.

Thus (8.18), in other words, specifies that segments are mapped (surjectively) onto phonological categories, preserving order.

An adequate graphic visualisation of a phonological representation combining (8.14) and (8.15) in accordance with principle (8.18) would in fact require three dimensions. Using dotted lines to link the tonal segments to the phonological categories, however, makes it possible to visualise such a structure, remembering that segments are ordered only on their own level, i.e. that the tonal segments are ordered only with respect to the other tonal segments:

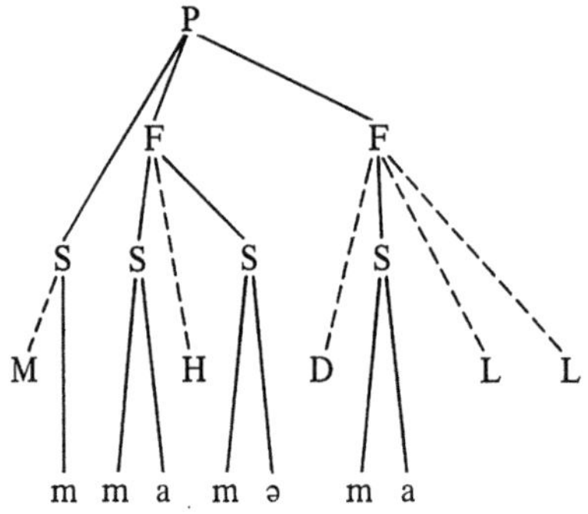

(8.19)

8.3.3 Underlying Tonal Representations

It now remains to be seen how a structure such as (8.19) is to be generated.

The occurrence of a (D) tone, just as is the case with many terraced-level tone languages [Welmers, 1959; Clements, 1976], is restricted in that it can only come after an (H) tone or another (D) tone. This sort of constraint is difficult to formulate other than by a transformational rule. It is generally assumed that a (D) tone is usually the surface reflex of an underlying sequence (L H). The classic example is from Akan [Fromkin, 1972], where

H L H
me ɔbo (8.20 a)

'my noun-prefix + stone'

surfaces as

H D
me bo (8.20 b)

A similar analysis has been proposed for English by Pierrehumbert [1980], who represents a downstepped sequence (H D D...) like

(8.21)

as a sequence of (H + L) tones. This, however, has unfortunate consequences in that it leads her to represent a sequence like

(8.22)

as a sequence of (H) tones, with a special rule of interpolation applying to create dips between adjacent (H) tones. Instead of this, it seems preferable to maintain a surface distinction between realisations like (8.21) and (8.22) but to assume that the two patterns are derived from the same underlying form by the optional application of a tonal assimilation rule converting a sequence

$$(X, L)_F \; (H \ldots)_F \quad (8.23\,a)$$

to

$$\underset{F}{(X)}\quad \underset{F}{(D\ldots)} \qquad (8.23\,b)$$

Such a rule is even more clearly assimilatory if we formulate it in terms of the feature analysis suggested above, since instead of (8.23) we have

$$_{F}\left(X, \begin{bmatrix}-\mathrm{Hi}\\+\mathrm{Lo}\end{bmatrix}\right)\quad {}_{F}\left(\begin{bmatrix}+\mathrm{Hi}\\-\mathrm{Lo}\end{bmatrix}\ldots\right) \qquad (8.24\,a)$$

converting to

$$_{F}X\quad {}_{F}\left(\begin{bmatrix}+\mathrm{Hi}\\+\mathrm{Lo}\end{bmatrix}\ldots\right) \qquad (8.24\,b)$$

This solution has the obvious advantage of allowing distinct representations for sequences like (8.21) and (8.22) while at the same time capturing the fact that the two patterns are basically very similar.

Adopting this analysis, we obtain as the underlying form for (8.19):

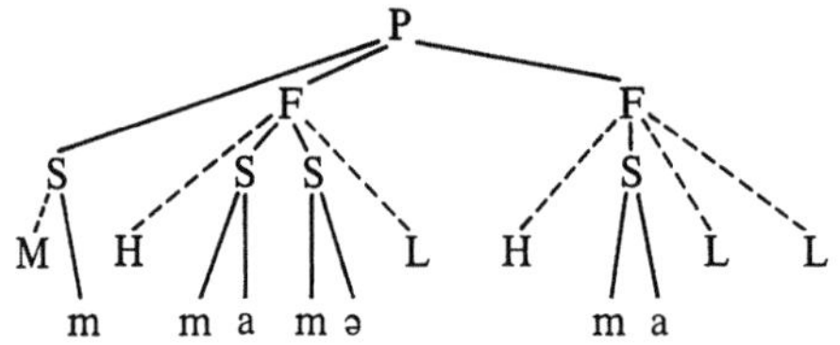

(8.25)

This, however, in turn, suggests a further generalisation. If we take the well-formedness condition (8.18) as a constraint on *surface* forms, as specifying in effect that phonological categories must be phonetically realised both segmentally and tonally, there is nothing to prevent us from assuming a slightly less linear underlying form for (8.25), such as:

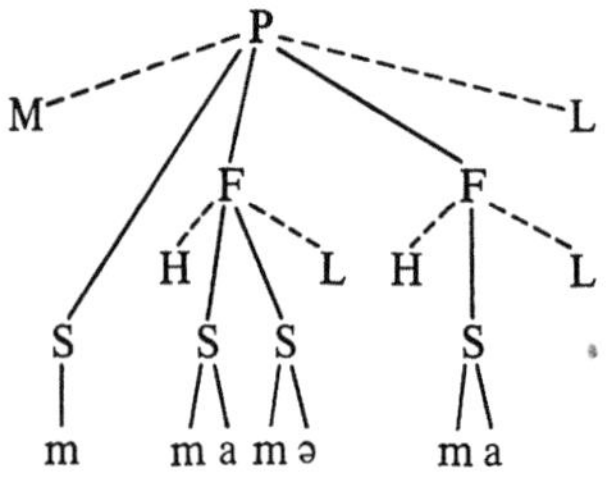

(8.26)

where each Foot dominates the tonal sequence (H L) and where the initial and final tones of the phrase are dominated directly by the category Phrase[8].

[8] This is in many respects comparable to Pierrehumbert's [1980] "boundary tones", as well as to Gårding's "sentence component" [this volume] and Thorsen's "sentence intonation" [this volume].

An ideal description of English intonation might simply allow tonal segments to be freely assigned in the phonology within the limits of, for example, a phrasal template such as:

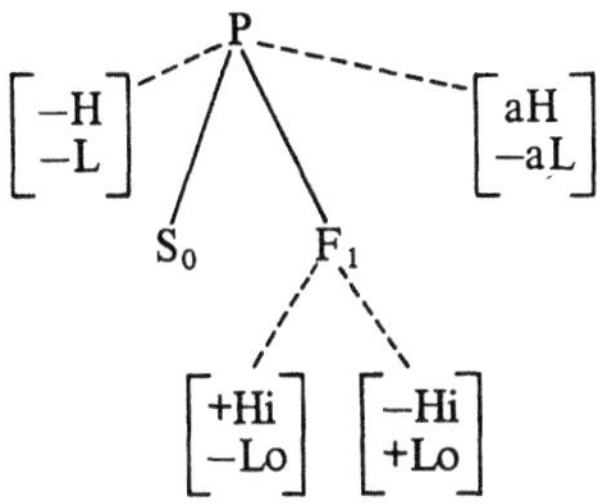

(8.27)

The sequences defined by (8.27) together with a floating (H) tone, part of the phonetic manifestation of emphasis in English, make up a core system of intonation for British English, a possibility which I explore in more detail in Hirst [in press].

8.3.4 Phonetic Implementation

Having worked our way up from the acoustic signal to a suggested underlying form, we can now work through a sample derivation. A sentence such as

A nice cup of coffee (8.28)

is provided with the following structure by the phonological component:

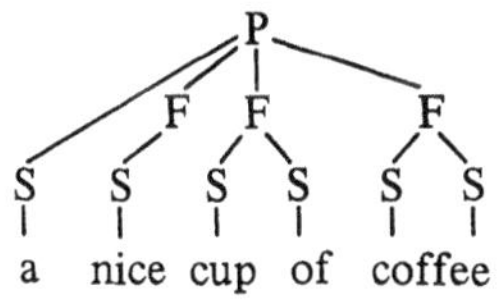

(8.29 a)

Tonal segments are assigned to (8.29 a) in conformity with the template (8.27), for example:

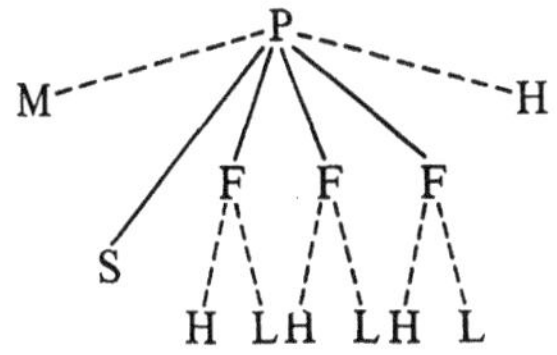

(8.29 b)

The downstep rule (8.23) will now apply, giving:

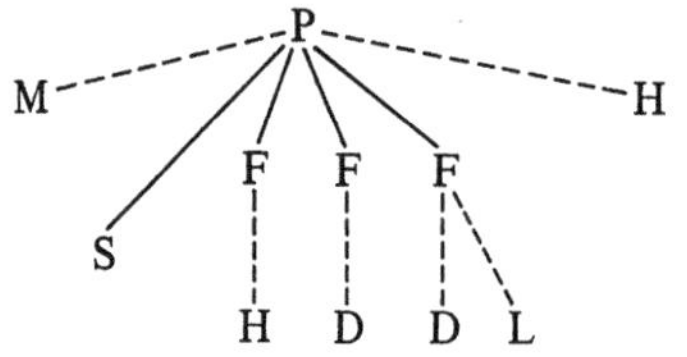

(8.29 c)

This in turn will be modified to conform to the surface constraint (8.18), so that the actual input to the phonetic component could be something like:

[[ə M]][nais H][kʌpəv D][kɔfi D L H]] (8.29 d)

The essential claim of such a theory of prosodic representations is that the tonal line and the non-tonal line are, fundamentally, independent. The duration of the various stress feet will naturally be determined partly on the basis of their various constituents on both the prosodic and the segmental level, but once this duration has been established, it is as if the rules determining the timing of the target pitches could not see beyond the foot structure of the sentence. Thus it is perfectly possible, for example, for a target pitch to be situated in the middle of an unvoiced consonant, a situation which does not facilitate the phonetician's task of estimating the value of the target pitch.

Without going into details concerning the actual F_0 values for a sentence like (8.29), we can represent the relative height of the target pitches as follows:

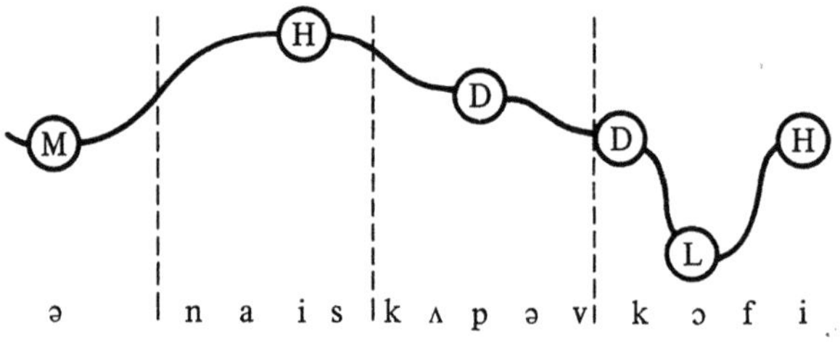

(8.30)

The continuous line represents the (spline) function interpolating between the target values. The actual F_0 values measured will be the result of the interaction between the underlying curve illustrated in (8.30) (modelling the evolution of vocal-fold tension) and the specific characteristics of the individual segments (voiced/unvoiced, occlusive, etc.).

8.4 Conclusion

It is evident from the above discussion that there is a complex interaction between choices at various levels of description. Thus the choice of phonological model will influence and be influenced by the particular phonetic model used to produce an actual F_0 curve. Hence the necessity of close cooperation between linguists and phoneticians in the description of such complex phenomena. In the same way, I suggested above that an adequate model for phonological representation might be a necessary prerequisite for a solution to the long-standing problem of how intonation contributes to the meaning of an utterance. I shall not attempt to explore this extremely complex question in any detail here; the following remarks are intended simply as programmatic suggestions.

If we assume that a grammar is organised as in (8.1) (repeated here):

$$\begin{array}{ccc} & \text{syntax} & \\ \swarrow & & \searrow \\ \text{phonology} & & \text{semantics} \end{array} \qquad (8.31)$$

we commit ourselves to the view that options in the phonological representation cannot be interpreted by the semantic component. According to the model developed in Sect. 8.3 above, however, the intonation of a sentence can be factored out into two component parts corresponding to

the phonological structure (8.32 a)
the underlying tonal sequence (8.32 b)

I also mentioned briefly the possibility that an emphatic morpheme exists in English which is phonetically realised as a floating (H) tone. This morpheme will naturally be present in the syntactic representation and will consequently be available for semantic interpretation. It also seems probable that the phonological structure is determined at least partly by the syntactic structure. This means that we can expect certain syntactic differences to be translated into distinct phonological structures. If, however, as I suggested, the tonal sequence is assigned freely in the phonology, this would imply that meaning differences attributed as a function of such tonal variations cannot correspond to semantic or syntactic differences, but must rather be explained as different pragmatic readings.

Rather than seek a purely syntactic, semantic or pragmatic explanation for the contribution of intonation to meaning, then, a more modular approach such as that proposed here might provide the key to the complex way in which intonation interacts with the other components.

9. Temporal Predictability in the Perception of English Speech

Hilary Buxton

The purpose of this contribution is threefold: (1) to review psychological studies of temporal structure in the perception of English speech, (2) to present some new experimental evidence which suggests that the perceptually crucial feature of temporal structure is its predictability, and (3) to discuss recent linguistic models of prosodic organisation in the light of that evidence.

9.1 Stress and Rhythm

English has sometimes been termed a "stress-timed" language. This label refers to a tendency of English utterances to equalise the temporal intervals between successive stresses, usually by compressing the duration of unstressed syllables. Thus the word "some" ought to be longer in (9.1), in which it forms the only unstressed syllable between two stresses, than in (9.2), in which it shares the interstress position with two other syllables:

John bought some balls (9.1)

John purchased some balloons (9.2)

Similarly, although any word uttered in isolation will be pronounced with at least one stressed syllable, whether this potential for stress is realised in a connected utterance can depend on the rhythmic context; thus "pretty" and "wanted" are stressed in (9.3) but not in (9.4):

She wanted some pretty parrots (9.3)

My aunt wanted ten pretty parrots (9.4)

Rhythmicity in English appears to be psychologically real both to speaker and to hearer. For instance, when slips of the tongue occur in which syllables are deleted or inserted, the resulting utterance tends to have a more equal number of syllables between stresses than the intended utterance [Cutler, 1980a]. Similarly, listeners who are asked to adjust a sequence of tones to match the rhythm of the sequence of stressed vowels in an English utterance tend to make the tone sequence rhythm more regular than the utterance rhythm actually was [Donovan and Darwin, 1979].

In fact, the physical basis for this supposedly isochronous rhythm is very slight (see, for instance, Bolinger [1965]); interstress intervals in actual utterances can vary by as much as 6:1. Nevertheless, recent work has indicated that physical isochrony may indeed be found if more subtle measurement techniques are employed. For instance, Hill et al. [1978] demonstrated a general tendency to isochrony in English speech, wherein, however, some rhythmic units are more isochronous than others: specifically, those that are

(a) utterance-final or
(b) include the main sentence accent.

Interestingly, psychological studies of the role of prosodic structure in sentence understanding have provided considerable evidence that both utterance-final and accented portions of the sentence may be easier to perceive. Most of these studies – and, indeed, three of the four studies to be reported below – used monitoring tasks, in which listeners are asked to respond as quickly as possible to the occurrence of a specific target (a word or a word-initial phoneme) embedded in a sentence. In phoneme monitoring, for example, targets at the end of an utterance are responded to faster than earlier targets [Foss, 1969], and targets on accented words are responded to faster than targets on unaccented words [Cutler and Foss, 1977]. The latter is perhaps not surprising, since there is evidence [Lieberman, 1963] that stressed words in running speech are simply acoustically more distinct than unstressed words. Yet the results of a study by Shields et al. [1974] cast doubt on a simple acoustically based explanation. Their study compared response times to the initial phoneme of nonsense words such as "benkik" where stress fell on the first or the second syllable. When the nonsense words were embedded in running speech, response times to targets on stressed syllables were faster than to targets on unstressed syllables; however, this difference disappeared when the same items were embedded in lists of other nonsense words. Similarly, Cutler [1976] found that phoneme-monitoring reaction times to targets which occurred in a position where the preceding prosody predicted that sentence accent would fall were faster than to targets which occurred where low stress was predicted, even when tape splicing had ensured that the target-bearing words were actually acoustically identical irrespective of position.

In addition, monitoring studies have shown the *disruption* of timing to have large effects on speech processing. For instance, Meltzer et al. [1976] showed that when words containing phoneme targets were slightly displaced from their position in normal speech, listeners' response times were significantly increased. More recently, Martin [1979] lengthened or shortened one vowel in a nonsense sequence to achieve a change in perceived tempo, and found that both lengthening and shortening increased reaction time to phoneme targets. The disadvantage was more pronounced for manipulations late in the sequence, suggesting that rhythmic expectancies are greater further into the utterance; and the effects persisted despite practice and familiarity with individual manipulated sentences, suggesting that a particular sentence has a rhythmic structure that must be kept intact for the sentence to be most efficiently perceived.

Of course manipulation of timing in natural speech will affect coarticulatory cues if the disturbance is close to the target. Martin and Bunnell [1981] found that anticipatory coarticulation was in fact not limited to immediately adjacent segments, though disruption of nearby coarticulatory cues resulted in larger effects than disruption of more distant cues. When timing was disturbed by the introduction of silent intervals immediately before the target [Meltzer et al., 1976; Martin et al., 1980], coarticulatory information available in advance allowed the target to be anticipated across the silent interval. This suggests that coarticulatory cues can provide information about "what" segment is coming up, independently of close temporal integration. Thus coarticulation effects can compensate in part for the disruption of speech timing.

Other evidence, however, suggests that regular versus disrupted timing is by no means the crucial issue in assessing the role of temporal structure in speech understanding. Some of this evidence will be described in the next section; first, however, it is necessary to introduce a crucial concept in what follows, namely the concept of a "P-centre" or syllable beat as a rhythmic alignment point.

9.2 P-Centres and Rhythm

Much previous research has investigated the perceived location of stress beats in sentences. For instance, Allen [1972] found that perceived stress beats were closely associated with the onsets of the nuclear vowels of stressed syllables. Rapp [1971] investigated production timing by measuring when nonsense words containing consonant clusters occurred with respect to a series of regular clicks when subjects were asked to produce the words in time with the clicks. The position of the click relative to the vowel onset correlated with the duration of the consonant cluster, showing not only that the vowel onset was an important factor in the relative timing of events, but also that the alignment was affected by the preceding acoustic waveform. These findings suggested that features of the acoustic waveform determine the location of the syllable beat, i.e. the rhythmic alignment point for the syllable.

Morton et al. [1976] discovered that when strings of isolated words were aligned such that their acoustic onsets occurred in a strict regular rhythm, the strings were not in fact *perceived* as regular at all. They proposed that rhythmic regularity is instead a function of the alignment of the "P-centres" (perception or production centres) of syllables. The P-centre of an isolated monosyllabic word corresponds to a syllable beat and can be thought of as the psychological moment of occurrence of the word. P-centres can be found for isolated monosyllables by allowing subjects to adjust the interval between two alternating items until perceptual regularity is achieved. If the acoustic onset of one of the items is taken as its P-centre, the P-centre of the other item is a function of its offset relative to its acoustic onset.

Marcus [1976], for instance, required subjects to control the alignment of alternating words until perceptual regularity was achieved; he then determined the P-centres relative to the onset of each word. He found that both initial consonant duration and subsequent vowel and consonant duration had an effect on P-centre location. Over 90% of the variance in P-centre placement could be accounted for by the algorithm: P-centre = a · (acoustic onset to vowel onset duration) + b · (vowel onset to acoustic offset duration) + c, where a is a parameter estimated at approximately 0.65, b is approximately 0.25 and c is a constant which is not assigned a value. The P-centre, then, is determined to a large extent by the acoustic waveform, but it is based on the entire stimulus rather than corresponding uniquely with any one feature of it.

The experimental work reported below relies on this concept of a P-centre as the rhythmic alignment point of a syllable or monosyllabic word. A series of experiments in which subjects were required to tap out the rhythms of natural continuous speech has been reported elsewhere [Buxton, unpublished]. The results of these experiments support the proposal that the pattern of P-centres marks the rhythm of an utterance in continuous speech. The two experiments reported in this section address the hypothesis that P-centre-aligned rhythms are *temporally highly predictable* and allow more efficient speech perception than non-aligned rhythms.

In the first study, subjects were required to press a response key as quickly as possible on detecting a target digit in a randomised list of digits. Two experimental tapes were made, containing lists of digits which were either onset aligned or P-centre aligned. Since the P-centre is the perceptual rhythmic alignment point, only the P-centre-aligned tape should sound completely regular, and the hypothesis tested was therefore that increased temporal predictability for this tape would result in faster reaction times. The digit lists were compiled from single exemplars of each of the digits one to nine using digital speech-editing facilities. Onset, offset and vowel onset were measured, and a P-centre for each word was estimated using Marcus's formula.

The type of alignment had a very strong effect on perception. The average reaction time in the onset-aligned condition was 629 ms, but in the P-centre-aligned condition only 459 ms. Serial position also had an effect: RTs to targets later in the lists were much faster than RTs to earlier targets. However, there was no interaction between serial position and alignment condition. (Even the first item of each list was predictable, since the lists were always preceded by a warning tone at a fixed interval before the first onset or P-centre.)

The 170 ms RT advantage for P-centre-aligned sequences in comparison with onset-aligned sequences must be due to some predictive process in perception. Note that in one sense the two tapes have the same temporal predictability, since for each digit there is an invariant timing rule. Nevertheless, physical regularity and perceptual regularity obviously do not coincide. Our perceptual processes appear to be sensitive only to the P-centre as a rhythmic alignment point, not to the onset of the word.

The next experiment investigated whether subjects could track excursions from a regular rhythm using a tapping task. Previous experiments [Buxton, unpublished] had suggested that subjects' tapping produces rhythms which correlate quite highly with predictions from P-centre algorithms. The present experiment used sequences of monosyllabic words which were either regular (i.e. had their P-centres aligned) or were irregular (i.e. had had some words displaced 100 ms from the regular timing). The monosyllabic words were excised from a recording of fluent, well-articulated speech using digital speech-editing facilities, and P-centres for the isolated words were calculated using Marcus's formula. The sequences formed simple sentences such as "A kind girl is here now". Subjects heard each sequence 12 times in succession; they were required to listen to the first 2 productions and then tap out the rhythm of the next 10 repeats.

It was obvious even during the running of the irregular tape experiment that the subjects were having great difficulty in tapping the rhythm; sometimes they failed to complete the sequences. Although only complete sequences were used in analysing the results, the variance of the data from the subjects listening to the irregular sequences was too high to allow comparison of the means with those listening to the regular tape. Therefore a comparison of the standard deviations was used to assess the effects of timing. The standard deviation of the intertap times averaged 30 ms in the regular condition but 55 ms in the irregular condition, a highly significant difference. In short, the sequences in the irregular condition did not have a rhythm that could be reliably tapped out even when exactly repeated 10 times through, which suggests that the temporal pattern of irregular sequences simply cannot be learnt. This result is consistent with the point made earlier concerning the data from Martin [1979], where effects of "tempo" change persisted despite practice and familiarity with individual sequences.

Arbitrary timing, then, is difficult to perceive, whereas P-centre-aligned timing is easy to perceive. However, since words do not occur with equally spaced P-centres in natural speech, it is somewhat paradoxical that the rhythm of natural utterances is also easy to perceive. For example, Buxton [unpublished] found that the typical standard deviation of intertap intervals for natural utterance rhythms was the same as that for P-centre-aligned rhythms, which suggests that the timing of natural speech has the same sort of inbuilt predictability.

The serial position effects found in monitoring tasks also attest to effects of temporal predictability. It might be argued that faster responses to later targets are in part determined by the prosodic effects described above, since in the default case sentence accent will occur towards the end of a sentence; yet the study by Shields et al. [1974] referred to in the previous section controlled the position of sentence accent, and still found an independent effect of serial position. Again, it might be argued that serial position effects are simply a reflection of the increasing probability of a target occurring, given that one must occur before the end of the sentence or list. Yet serial position effects also occur in experiments in which a large number of sequences contain no targets at all. The next two experiments investigated the temporal predictability of continuous speech and its effects upon monitoring performance.

9.3 Rhythm and Temporal Predictability

Both of the experiments to be described in this section measured listeners' response times to word-initial phoneme targets which occurred in fluent, rhythmic natural speech. In the first experiment, the natural speech was compared with three other kinds of speech material:

1) Speech which was prosodically intact but semantic nonsense: the "jabberwocky" condition. Jabberwocky sentences were constructed from natural sentences by replacing all initial consonant clusters – e.g. "You are a sweet girl I know so why do it" became "Wou are a gleet dirl I fow jo dy ku it". The jabberwocky sentences were read with the timing and intonation of the originals.

2) Speech which was in some sense syntactically and semantically intact, but in which the prosody was disrupted: the "concatenated" condition. Concatenated sentences consisted of words excised from fluent, well-articulated sentences and strung together to form the same set of sentences used in the natural condition.

3) Speech in which both prosody and syntax/semantics were disrupted: the "random" condition. These sentences were constructed in the same manner as the concatenated sentences, except that the words were strung together in random order.

Phoneme targets in each of the four conditions could be either stressed or unstressed, and could occur in an early, middle or late position in the utterance.

Briefly, there were main effects for all manipulations:

(a) response time to targets was fastest in natural speech and slowest in the random condition in which both prosody and semantic/syntactic structure were disrupted, with the other two conditions falling in between;

(b) stressed targets were responded to faster than unstressed ones; and
(c) the usual serial position effect was observed.
Of greater interest, however, were the interactions. Most importantly for our argument, there was a significantly greater serial position effect for both kinds of rhythmic material (normal and jabberwocky) than for either kind of arhythmic speech (concatenated or random). Exactly the same interaction held for the stress effect: it was significantly greater in rhythmic than in arhythmic material. Moreover, it also seems that the rhythm gives particular advantage to stressed targets in middle and late positions. Consider the matched sentences of the natural and concatenated speech. The absence of prosody does not seem to affect the early targets, nor particularly the unstressed targets. The later stressed targets are, however, responded to faster. Mean reaction times (in ms) are given below.

	Serial position		
	Early	Middle	Late
Unstressed			
Natural	693	598	575
Concatenated	692	650	595
Stressed			
Natural	604	487	428
Concatenated	587	562	504

This finding is consistent with that of Shields et al. [1974], who found that stressed targets were responded to faster than unstressed targets in sentence context, but that when the target items were excised and put in a list, the advantage of the stressed targets was lost. In other words, prior rhythmic context seems to be necessary for faster reaction times in natural speech.

In the second experiment, rhythmic expectations were investigated using cross-spliced sentences in which the number of syllables in a noun phrase was manipulated. It was assumed that the timing of the material that preceded this noun phrase would be affected by the rhythmic integration of the complete utterance, i.e. that it would take into account the number of syllables in the noun phrase. If hearers could extract this information from the relative timing of the utterance, this could lead to faster reaction times when the noun phrase was matched to its context than when it had been spliced from a different context. To avoid any confounding of coarticulation effects, all experimental material was cross spliced, either with rhythmically matched or unmatched utterances. An example sentence is given in (9.5).

target /t/	(the cross-splice point is marked //)	(9.5)
"short context"	"The small girl was playing with the // red toy in the garden of the largest house"	(9.5 a)
"long context"	"The small girl was playing with the // reddish toy in the garden of the largest house"	(9.5 b)
"short target"	"The small girl was playing with the // blue toy in the garden of the largest house"	(9.5 c)
"long target"	"The small girl was playing with the // bluish toy in the garden of the largest house"	(9.5 d)

The four original sentences were cross spliced to produce four experimental sentences; the first half of each of the context sentences was spliced onto the second half of each of the target sentences. The target phoneme always began the noun of the noun phrase following the cross-splice point – in (9.5), the /t/ of "toy". The cross splicing did not cause any obvious distortion to the timing of the sentences.

There was no main effect either of "long" versus "short" words preceding the target, or of original context, but exactly the predicted interaction was found: targets preceded by "long" words were responded to significantly faster when the context had also contained a "long" word, whereas targets preceded by "short" words were responded to faster when the context had contained a "short" word. In (9.5), that is, response times to /t/preceded by "blue" were faster when the context had come from version (9.5 a) of the sentence, containing the word "red", than when the context had come from (9.5 b), with "reddish", whereas response times to the /t/ preceded by "bluish" were faster with context (9.5 b) than with context (9.5 a).

There was also a significant effect of serial position in this experiment. However, the effect was not in this case greater for the rhythmically better-integrated sentences than for the less rhythmic ones. It should be noted, though, that the effect of rhythmicity in this experiment, though significant, was also much smaller than in the previous experiment, so that perhaps any differential effects of rhythmicity across serial position were simply too small to be measured with this kind of material.

In conclusion, however, the experiment has clearly shown how potent rhythmic expectancies can be in continuous speech. The "short" rhythmic context set up a temporal expectancy which matched the "short" target sentence better than the "long" sentence, and this matching difference was reflected in the subjects' response times. Conversely, the "long" context predicted the timing of the "long" target sentence better than that of the "short" target sentence, and again the predictable rhythm produced faster response times than the unpredictable. Clearly listeners are making good use of the temporal predictability of natural speech in understanding.

9.4 Modelling Rhythmic Structure

The evidence discussed so far points to the existence in natural speech of a temporal structure which can be used by listeners to anticipate stress pattern. Until recently, however, descriptions of this structure have been limited to approximate claims about a "tendency to isochrony" [Hill et al., 1978]. Most synthetic speech systems simply have a complex set of rules based on features of individual words or segments that allow an approximation to the timing of segment durations found in natural speech; such an approximation fails to capture the rhythmic integration of whole phrases that appears to be characteristic of fluent utterances.

Since the publication of Liberman and Prince's influential article [1977], however, linguists have devoted increasing attention to what has been called metrical structure in language. In order to explain a variety of rhythmic phenomena in stress assignment, Liberman and Prince proposed that phonological elements are grouped into hierarchical constituents in a binary branching tree with the relation strong/weak (s/w) defined on the pairs of nodes, e.g.

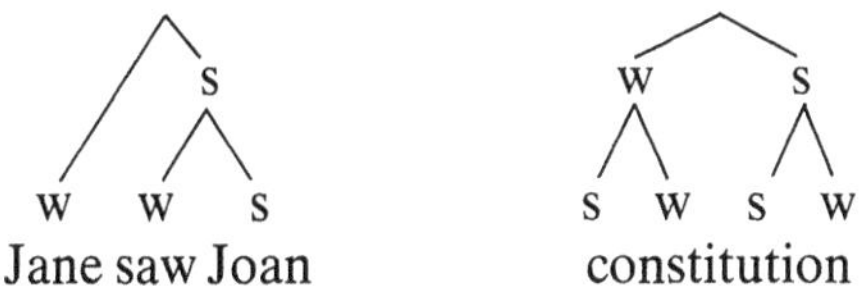

Positing such a structure makes it relatively easy to account for the alternation of stressed and unstressed syllables in a word or phrase, and for such phenomena as the reversal of stress seen in the pair "thirtéen" : "thírteen mén". Modifications of Liberman and Prince's proposals by Selkirk [1980] make it possible to account as well for the difference seen in the less prominent syllables in, for example, "modest" versus "gymnast", within the context of a fairly unified notion of prosodic structure.

However, the major concern of work based on the proposals of Liberman, Prince and Selkirk has been with the issues of word stress assignment [e.g. Hayes, 1981] and prosodically influenced syntactic phenomena [e.g. McCarthy, 1979; Nespor and Vogel, this volume]. Very little has been said about the question of temporal predictability discussed here, the most obvious exception being the notion of the Metrical Grid. In its original conception [Liberman, 1975], this was intended to model the temporal structure of actual utterances, but in Liberman and Prince [1977] and in subsequent work [e.g. Thompson, 1980] it is little more than an algorithm for computing levels of stress. Since Liberman's original notion has not been further developed, it is difficult to assess its adequacy as a basis for explaining temporal predictability. However, one general comment can be made.

Predictive processes in speech perception operate in real time [Marslen-Wilson and Welsh, 1978]. Perceptual effects of semantic predictability, for instance, are found well before the end of the phrase, and indeed, responses showing the influence of predictability are so fast that they may even occur before the end of the target word. A strictly hierarchical model of prosodic structure such as the Liberman-Prince-Selkirk model, however, requires that the end of the unit be known before the structure can be specified; this applies a fortiori to the Metrical Grid, which is derived from the hierarchical branching structure. This in turn renders it unlikely that real-time predictability can be modelled by the Metrical Grid in its present form.

At the same time, however, the very existence of a phonological model that allows for the possibility of a complex prosodic structure appears to be an advance over merely linear models of stress and timing. On the basis of the experimental data presented here, it would seem likely that future research on the pattern of P-centres in continuous speech will allow an exploration of the way in which rhythmic structures of the language are realized in the production of an utterance. Understanding the encoding of information in the rhythmic structures themselves, though, requires an appropriate phonological description. We need input from both linguistic and psychological research on the rhythmic structure of language for a full model of speech production and perception.

9.5 Summary

The four experiments discussed in this contribution investigated the effects of temporal predictability on speech-monitoring tasks. First, P-centre-aligned (perceptually regular) lists of digits gave faster target response times than matched onset-aligned (perceptually irregular) lists. This result suggests that temporal predictability is relevant for perceptual processing. Second, subjects could reliably tap out the rhythm of perceptually regular lists, but gave very erratic performances when required to tap out the rhythm of irregularly timed lists, even after several repetitions. Arbitrary temporal patterns are not easy to learn. Since the rhythm of unequally spaced beats in natural utterances can be reliably tapped out, there appears to be a predictable temporal structure in natural speech. Third, rhythmic structuring of speech material was found to enhance both the serial position effect and the stressed item advantage in a phoneme-monitoring task. Fourth, an effect of rhythmic expectancy on phoneme-monitoring time was found when the number of syllables in a noun phrase was manipulated in cross-spliced sentences. Faster reaction times were found when the context matched the original noun phrase than when it did not. This result suggests that rhythmic expectancies from prior temporal context affect perception.

These results are interpreted as supporting the view that English utterances have a rhythmic structure that can give greater temporal predictability to stressed than to unstressed items and that later items in an utterance are more temporally predictable than earlier items. Speech perception involves predictive processes operating on the temporal structure as well as on the semantic context. Any functional model put forward to account for speech production and perception must therefore encode stress information in the rhythmic structure and incorporate predictive processes for its extraction.

Acknowledgement. I would like to thank Anne Cutler, Bob Ladd and John Morton for valuable contributions to this paper. The research described in it was carried out under the supervision of John Morton at the Applied Psychology Unit, Cambridge, while the author held a Medical Research Council studentship.

10. Prosodic Structure Above the Word

Marina Nespor and Irene Vogel

10.1 Introduction

Since in various studies the term "prosody" has been used in a variety of ways, we will begin by clarifying our use of it in this contribution. We include under "prosodic phenomena" any phonological rules or processes that are not purely local, in that they cannot be described solely in terms of their phonotactic environments. Instead, additional information is required as to what larger units, or "prosodic domains", they belong to. In other words, while certain phonological phenomena can be described with rules that modify particular segments in particular segmental contexts, these segmental contexts are not enough to determine whether or not a rule applies. Following recent proposals [e.g. Liberman and Prince, 1977; Selkirk, 1978 b, 1980], we take the prosodic domains to include rhyme, syllable, foot, phonological word, phonological phrase, intonational phrase, and utterance. To take a simple illustration, consider a rule in Dutch that inserts a schwa between a liquid and a following consonant[1].

$$\phi \rightarrow \text{ə}/\text{L}\_\_\text{C} \qquad (10.1\,\text{a})$$

For example, *melk* "milk" can be pronounced *mel[ə]k*. A schwa cannot be inserted, however, in the identical segmental context in the word *balkon* "balcony". The difference between these two cases is that in the first one the liquid and consonant are in the same syllable, while in the second one they are not. Thus, the rule for schwa insertion must include information about the prosodic domain in which it applies, in addition to the segmental specifications, as in the revised formulation in (10.1 b).

$$\phi \rightarrow \text{ə}/[_{\sigma}\ldots\text{L}\_\_\text{C}\ldots]_{\sigma} \qquad (10.1\,\text{b})$$

In this contribution, we will be concerned with phenomena whose domains are larger than the phonological word, that is, external sandhi rules

[1] The extent of application of this rule is subject to the influence of such factors as style of speech and social class, as well as the nature of the consonant following the liquid, but these details are beyond the scope of this paper.

(ESR's) and intonation phenomena. Contrary to earlier proposals that the domains for ESR's are expressible in terms of syntactic constituency [Selkirk, 1972], it is our claim that the appropriate domains for ESR's are the prosodic categories of phonological phrase, intonational phrase and utterance, which, while they make use of syntactic information in their construction, are not necessarily isomorphic to syntactic constituents [Selkirk, 1978 b]. In the following sections, we will give rules to construct these prosodic categories and argue that in addition to accounting for the application of ESR's in a number of languages, our categories specify the domains of other phenomena commonly analyzed in the literature on prosody, such as constituent-final lengthening and pauses [Cooper, 1976; Lehiste, 1970; Cooper and Paccia-Cooper, 1980]. Finally we will show, on the basis of a perception test involving ambiguous sentences in Italian, that in a number of cases disambiguation is correctly predicted by our prosodic categories, while syntactic constituency makes incorrect predictions.

Our view of prosody fits into the current proposals regarding the organization of the grammar and, in particular, the phonology, within a generative-transformational framework. Specifically, we are assuming a model of grammar in which the syntactic component generates sentence structures which then serve as the input to a logical form component and a phonological component [Chomsky, 1981]; there is no direct interaction between the logical form and phonological components, a point we will return to below.

10.2 Construction of Prosodic Categories

On the basis of an analysis of two phenomena in Italian, the external sandhi rule *raddoppiamento sintattico* and the spreading of intonation contours, we have developed a proposal about the internal structure and prominence relations of the prosodic categories of phonological phrase and intonational phrase, respectively. The rules for constructing these categories and the additional prosodic category, utterance, are given below.

Phonological phrase (φ)
φ construction. Join into a φ any lexical head X with all items on its nonrecursive side within the maximal projection and with any other nonlexical items on the same side (e.g. prepositions, complementizers, conjunctions, copulas). (10.2 a)

φ constituency. φ branches in the same direction as the syntactic trees. (10.2 b)

Following Wheeler [1981], we assume that prominence relations within a tree are derived from the direction of branching. That is, right-branching trees are labelled weak/strong (w/s) and left-branching trees s/w. The combination of this assumption and our φ construction and constituency rules predicts that the head of the syntactic phrase on the basis of which a φ is formed is its most prominent element[2].

Examples of φ constituency are given in (10.3).

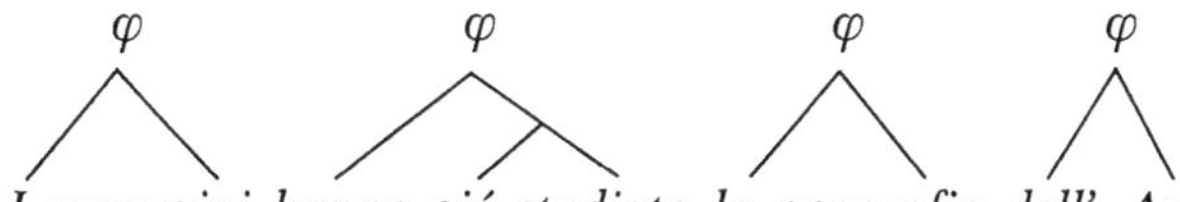

I ragazzini hanno giá studiato la geografia dell' Asia.

"The children have already studied Asian geography." (10.3 a)

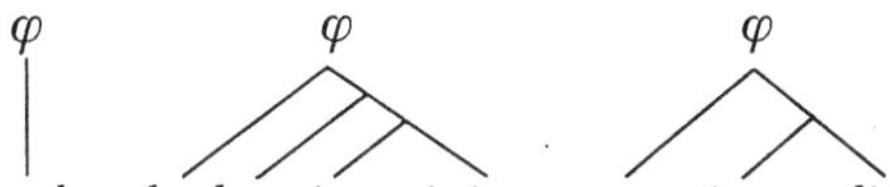

Credo che la mia opinione non sia gradita.

"I think my opinion isn't welcome." (10.3 b)

The rules for constructing the second category, the intonational phrase, and the last category, the utterance, are given in (10.4) and (10.5), respectively.

Intonational phrase (I)
I construction. (i) Any displaced syntactic constituents, parentheticals and nonrestrictive relative clauses obligatorily form an I. (ii) Starting with the first phonological phrase, join as many phonological phrases as possible into an I until either (a) the end of the maximal projection of an N is reached, or (b) another $\bar{S}$ begins. Once such an I is formed, proceed in the same way until the end of the main sentence is reached. (10.4 a)

I constituency. I is right branching. (10.4 b)

Utterance (U)
U construction. Join all I's in a root sentence into a U. (10.5 a)

U constituency. U is right branching. (10.5 b)

Given the assumption that right-branching trees are labelled w/s, by specifying that I and U are right branching, we assign to their rightmost elements the strongest prominence, thereby accounting for the observation that lengthening is characteristic of the ends of such prosodic units.

2 Since the labelling of s and w is not directly relevant to the issues discussed in this paper, these designations will not appear in the prosodic trees presented here.

Examples of I and U construction are given in (10.6).

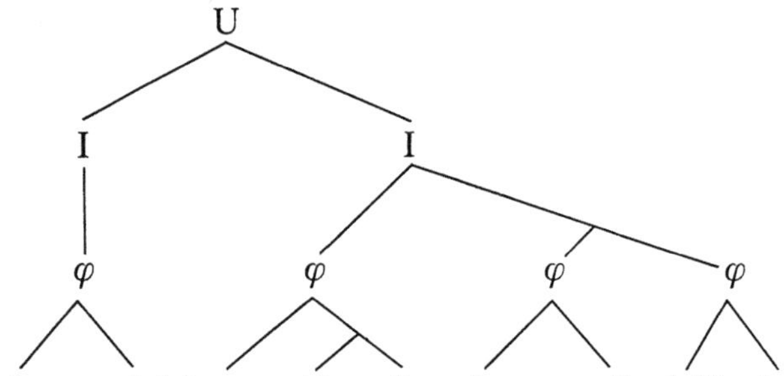

I ragazzini hanno giá studiato la geografia dell' Asia (10.6 a)

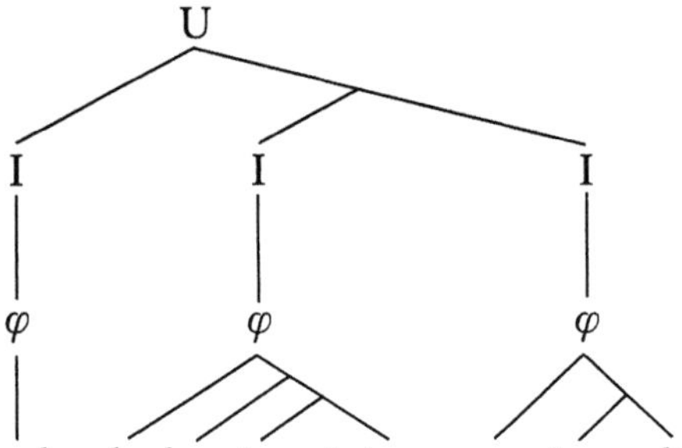

Credo che la mia opinione non sia gradita. (10.6 b)

In addition to these basic rules for constructing φ, I and U, there are a few optional restructuring rules which take into account linguistic as well as extralinguistic considerations, such as physiological limitations, perceptual strategies and stylistic factors. The phonological phrase can be optionally restructured according to the following rule, an illustration of which is given in (10.8).

φ-restructuring. A nonbranching φ which is the first complement of X on its recursive side loses its label and is joined to the φ containing X, under a new node labelled φ'. (10.7)

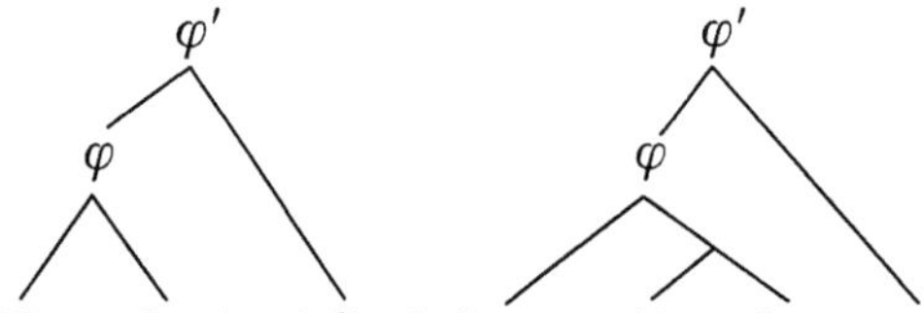

Gli studenti migliori hanno giá studiato tutto.

"The best students have already studied everything." (10.8 a)

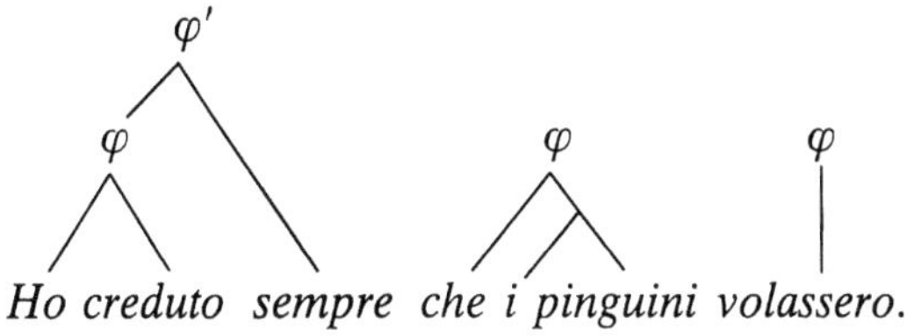

"I have always thought that penguins fly." (10.8 b)

For the category I, restructuring reflects the tendency to equalize the lengths of the individual units while still respecting linguistic structure. Thus, particularly short I's tend to be combined into larger I's, as shown in (10.9 a), and particularly long I's tend to be broken down into shorter I's, as shown in (10.9 b).

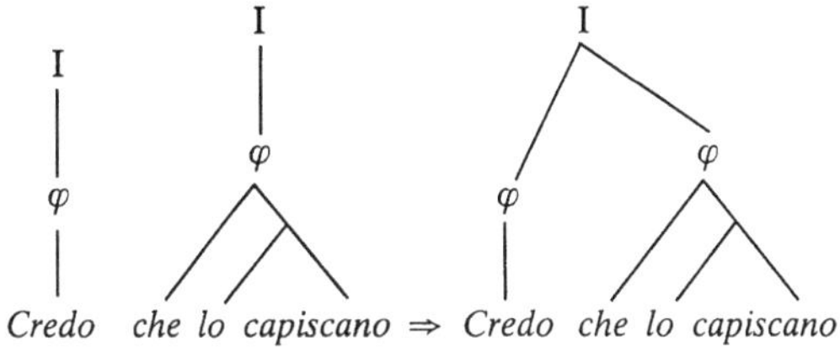

"I think they understand it." (10.9 a)

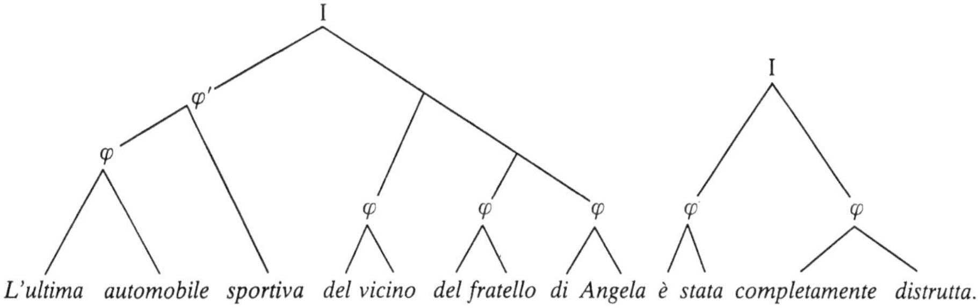

"Angela's brother's neighbor's latest sports car has been completely destroyed." (10.9 b)

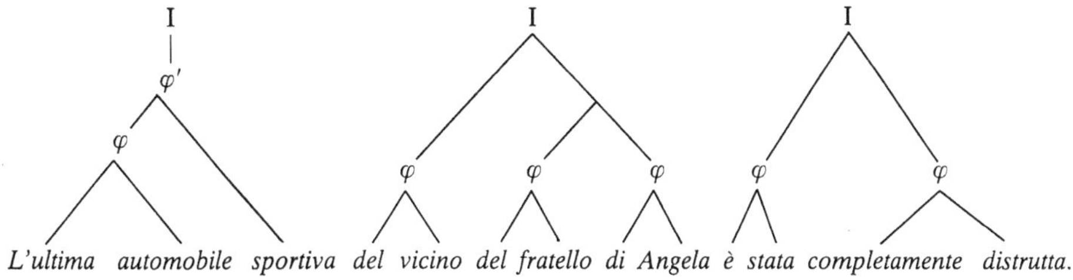

Even U can be restructured in very special cases, for example the conjoining of very short root sentences uttered in rapid succession, as illustrated in (10.10).

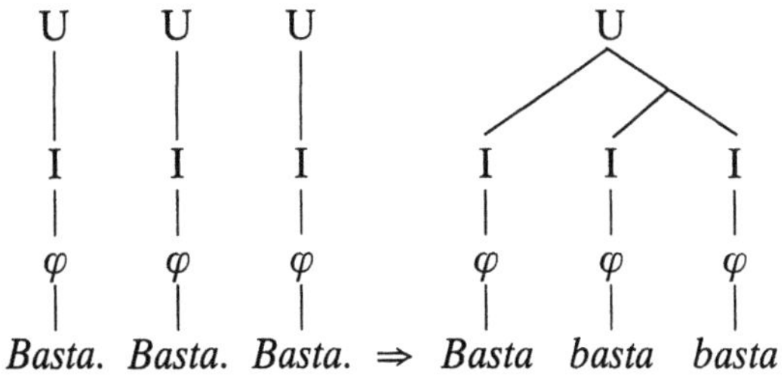

(10.10)

"Enough. Enough. Enough."

We have demonstrated elsewhere [Nespor and Vogel, 1982] that the prosodic categories just illustrated are capable of accounting for the contexts of application of several ESR's of Italian and other languages, while we have found no ESR that applies in different domains. We will illustrate here the functioning of only one such ESR, *raddoppiamento sintattico* (RS).

RS is a rule of central and southern varieties of standard Italian that results in the lengthening of the initial consonant of $word_2$ in a sequence $word_1$ $word_2$, under certain phonological and syntactic conditions. The functioning of RS is exemplified in (10.11).

Avrá [p:]arlato. "He will have spoken." (10.11 a)
Avéva [p]arlato. "He had spoken." (10.11 b)

Notice that the [p] is lengthened only in (10.11 a), that is, following a stressed vowel at the end of $word_1$, and not in (10.11 b), where $word_1$ does not end in a stressed vowel. The phonological factor of stress is not, however, sufficient to determine the application of RS. While it is clear that RS does not apply after just any word that ends in a stressed vowel, predicting exactly where it applies is a problem which has puzzled scholars of Italian linguistics for years. To see how complicated this problem is, consider the examples in (10.12). In both sentences the same sequence of words – *perché Paolo* – appears, though RS applies only in (10.12 a).

Perché [p:] aolo non vuole crederci?
"Why doesn't Paolo want to believe it?" (10.12 a)
Che c'é un perché [p] aolo lo crede.
"Paolo believes that there is a reason." (10.12 b)

If we now consider the prosodic structure of the sentences in (10.12), we see that *perché* and *Paolo* form a phonological phrase in (10.12 a) but not in (10.12 b), as illustrated in (10.13).

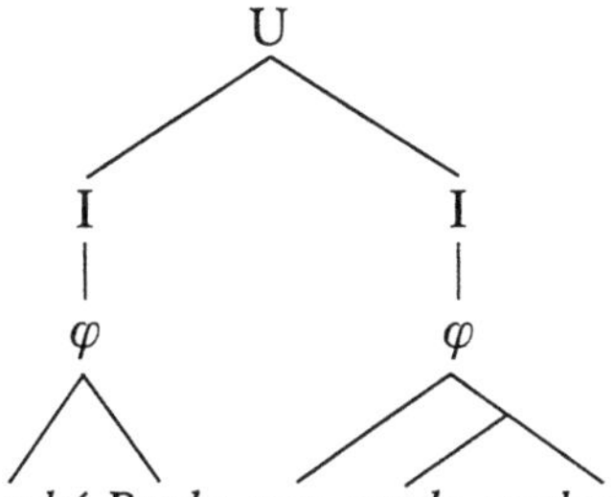

Perché Paolo non vuole crederci? (10.13 a)

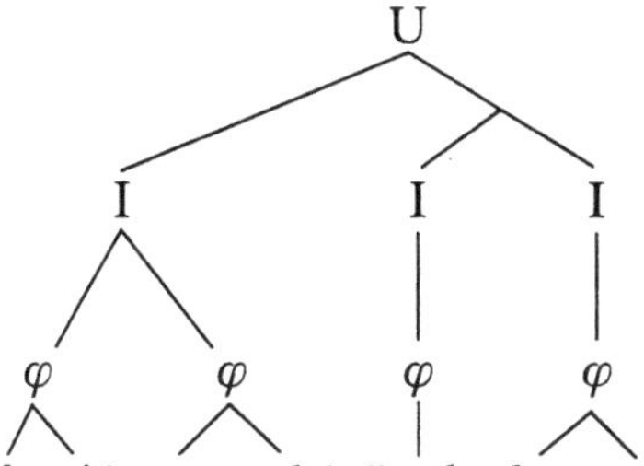

Che c'é un perché Paolo lo crede. (10.13 b)

While *perché* and *Paolo* are dominated by different I nodes in (10.13 b), and by the same I in (10.13 a), in (10.14) it can be seen that this is not what accounts for the presence or absence of RS.

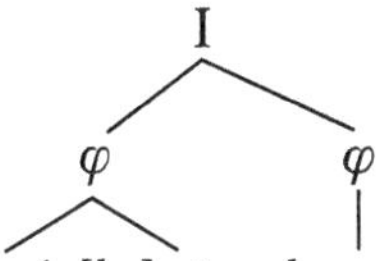

Sará [k:]otto domani.

"It will be cooked tomorrow." (10.14 a)

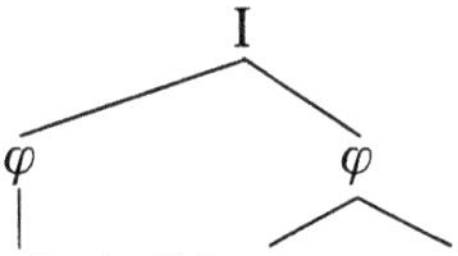

Venderá [k]astagne arroste.

"He will sell roasted chestnuts." (10.14 b)

In both sentences the words in question (i.e. *sará cotto* and *venderá castagne*) are dominated by a single I node, but RS applies only in the first sentence, that is, where the two words are members of the same φ. The phenomenon of RS thus shows how a prosodic category can provide a simple and straight forward way of specifying the domain of an external sandhi rule.

10.3 Intonation Phenomena

In addition to allowing us to specify the domains of external sandhi rules, the prosodic categories described in the preceding section make it possible to account for certain facts of intonation. In this section, we will concentrate on the intonational phrase and show that it enables us to explain a specific intonation phenomenon, namely the position of intonation breaks, which are manifested by lengthening of the elements preceding such a break, insertion of a pause, or the potential for inserting a pause. A comparison of the syntactic and prosodic treatments of this phenomenon will reveal that syntax does not explain a number of problems that prosodic structure does. In particular, we will examine two points of difference: intonation breaks before relative clauses and the flexibility of intonational phrases.

We will begin by considering intonation breaks before relative clauses, a phenomenon often discussed in the literature [Chomsky, 1965; Chomsky and Halle, 1968; Langendoen, 1975]. It has been noted that complex sentences containing restrictive relative clauses present a problem for intonation in that "the intonation breaks are ordinarily inserted in the wrong place" [Chomsky, 1965, p. 13]. Underlying Chomsky's statement is the assumption that if intonation breaks were in the "correct" place, they would reflect the surface syntactic structure of the sentence, in particular the direction of embedding. Compare the bracketings of the sentences in (10.15) and (10.16); those in (10.15) are predicted by the syntax and those in (10.16) correspond to the position of intonation breaks.

This is [the cat that caught [the rat that stole [the cheese]]] (10.15)
[This is the cat] [that caught the rat] [that stole the cheese] (10.16)

According to Chomsky and Halle [1968, p. 372], this discrepancy between how things are and how things should be is "obviously a matter of performance rather than of grammatical structure." The notion of phonological phrase is introduced, and the rules that insert the boundaries of such phrases "will have to take account of syntactic structure, but they will also involve certain parameters that relate to performance, e.g. speed of utterance."

It seems, however, that the location of intonation breaks in a sentence such as the one above is not a matter of performance, but rather of competence, since it is clearly rule governed and native speakers have strong intuitions about the location of such breaks. In addition, two arguments are given by Langendoen [1975] in favor of considering the rules that define intonation boundaries as part of the grammar. The first argument is that these rules are capable of introducing ambiguities, which is not a property of performance rules. The second argument is purely theory-internal and is based on the observation that a rule that would "readjust" the structure of (10.15) to give

(10.16) must, for formal reasons, be considered part of the grammar [Langendoen, 1975, p. 546].

Langendoen provides further arguments to show that while the readjustment rules must be considered part of the grammar, they cannot be part of the syntactic component. Since he excludes a priori the possibility that such rules are part of either the semantic or the phonological component, he concludes that there must be yet another component of the grammar: the readjustment rule component. It should be noted, however, that the only motivation for modifying the syntactic structure is phonological and, more specifically, prosodic. Given the recent developments in phonological theory mentioned above, which take into account prosodic as well as segmental phenomena, the observations made by Langendoen and others are automatically accounted for in the phonological component, thus eliminating the need for a separate readjustment component. In particular, the rules we have given above for I construction allow us to account correctly for the intonation breaks located at the end of I's in sentences such as (10.16). This is illustrated in (10.17), where the boundaries of the I's correspond to the positions where intonation breaks may be found.

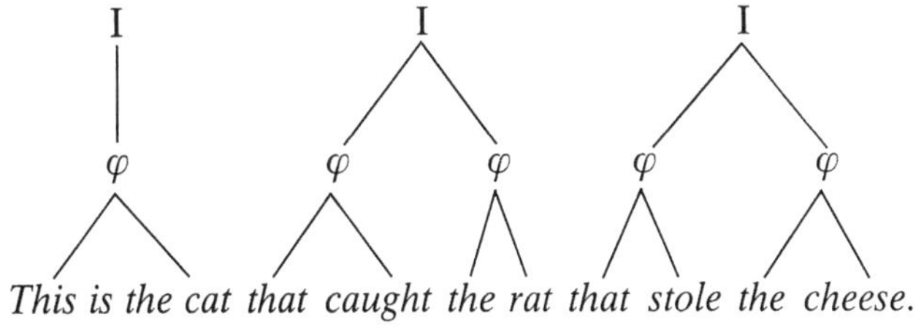

(10.17)

Notice that a general feature of the prosodic hierarchy is that it is flatter than the corresponding syntactic hierarchy. Due to the non-recursive nature of prosodic structures, we are limited to a depth of embedding of three levels above the word level, as opposed to the unlimited embedding possibilities offered by the syntax. As a result, certain relationships represented by syntactic structure are eliminated in prosodic structure, thus reducing the number and types of syntactic distinctions that can be reflected in the phonology. For example, while the three clauses in the sentence in (10.17) are all at different levels of embedding, their prosodic structures are the same.

The flexibility of prosodic categories, in particular I, permitted by the restructuring rules discussed in the preceding section, not only allows us to account for variability in the application of ESR's, but is also pertinent to the location of intonation breaks in a sentence. Since the prosodic category I can be restructured to eliminate either particularly long or particularly short constituents, the prediction is made that intonation breaks, which occur at the end of an I, may be distributed in different ways in the same sentence. For example, while it is possible to read the sentence in (10.18) with a pause after each I, as indicated by the bracketings in (10.18a), since these I's are particularly short, restructuring is likely to take place, yielding the pronunciation with intonation breaks indicated in (10.18b), or even a pronunciation without intonation breaks, as in (10.18c).

[The frog]$_I$[ate a fly]$_I$ [for lunch]$_I$ (10.18a)

[The frog]$_I$[ate a fly for lunch]$_I$ (10.18b)

[The frog ate a fly for lunch]$_I$ (10.18c)

Similary, while the sentence in (10.19) can be read with intonation breaks at the end of the I's, as indicated by the bracketings in (10.19a), since the I's are particularly long, it is likely that restructuring will take place, giving the pronunciation with intonation breaks indicated in (10.19b).

[That book of adventures of savage pirates of the dangerous days of yore]$_I$ [frightens children of all ages]$_I$ (10.19a)

[That book of the adventures of savage pirates]$_I$ [of the dangerous days of yore]$_I$ [frightens children of all ages]$_I$ (10.19b)

The fact that syntactic constituency provides only one way of grouping the words of a given sentence predicts that intonation breaks must be distributed in one particular way for a given sentence, say at the ends of NP's and VP's [Cooper and Paccia-Cooper, 1980]. Thus the various pronunciations of the above sentences are more appropriately handled in terms of prosodic constituents. In the next section, we will examine another type of data for which prosodic constituents provide a more adequate account than do syntactic constituents.

10.4 Ambiguity and Potential for Disambiguation

While the proposal regarding prosodic structure presented here was originally developed to account for the application of external sandhi rules and the distribution of intonation contours, it turns out that the prosodic categories constructed by the rules given above also allow us to make interesting predictions regarding ambiguous sentences.

Sentences with identical sequences of words, or at least sound segments, but with different meanings, are termed ambiguous. It has been suggested [Lehiste, 1973; Cooper and Paccia-Cooper, 1980] that not all such sentences are in fact equally ambiguous. Rather, those sentences with different syntactic structures may be disambiguated on the basis of suprasegmental factors dependent on the syntax, while those that have the same syntactic structures cannot be disambiguated. In this paper, we have advanced the claim that prosodic and syntactic structures are not in a one-to-one relationship with each other, but instead differ in a number of ways. In particular, within a phrase, the prosodic category φ/φ' is capable of making distinctions that the syntax does not make, and furthermore , the prosodic category I eliminates differences based on the level of embedding that are crucial for the syntax.

In this section, we will examine a number of different types of ambiguous sentences with a variety of syntactic and prosodic structures in order to determine whether it is a difference in syntactic constituency or in prosodic constituency that leads to disambiguation. The different predictions made by syntactic and prosodic structures are evaluated on the basis of a perception test.

10.4.1 Syntactic Versus Prosodic Constituents, and the Resolution of Ambiguity

When two sentences have the same sequence of segments, a problem arises for the hearer in identifying the correct meaning, in particular when the contextual information is not adequate. In such cases, the hearer must make use of another type of information, namely prosodic information. The questions that arise are (from the speaker's side) where this prosodic information comes from, and (from the listener's side) how this information can be associated with different meanings. A commonly accepted hypothesis, as was mentioned above, is that there is a direct relationship between the syntactic structure of a sentence and its prosody. This hypothesis predicts that ambiguous sentences with different syntactic structures for the alternative readings will automatically differ prosodically, and will therefore be disambiguable. On the other hand, ambiguous sentences with identical syntactic structures will be identical prosodically and thus will not be disambiguable. We have proposed elsewhere that it is not the syntactic structure that determines whether the two readings of an ambiguous sentence are distinguishable, but rather the prosodic structure, which, while it makes use of syntactic information in its construction, differs in certain ways from the syntactic structure [Nespor and Vogel, in press].

The syntax of the two distinct readings of an ambiguous sentence can be the same or different on two independent dimensions: syntactic *constituency*, and syntactic *labels*. Prosodically, too, the two readings of an ambiguous sentence may also differ, or be the same, in their constituency. Since we are dealing with sentences, the prosodic categories that are relevant here are those above the word level, that is, φ and I. Once again, these two factors are independent of one another.

The prosodic and the syntactic dimensions are not independent, however, since the construction of prosodic constituents depends on syntactic structure. When both readings have an identical syntactic constituent structure, their prosodic structure must also be identical. However, in all other cases, φ and I constituency may vary freely with respect to the two syntactic factors. Thus there are ten possible relations which the two readings of an ambiguous sentence may bear to one another, as represented below (with the exception of category 2, the numbers refer to the rows in Table 10.1, see p. 138):

			Syntax			
			Constituents Same		Constituents Different	
			Labels Same	Labels Different	Labels Same	Labels Different
Prosody	φ Same	I Same	1	2	3	4
		I Different			5	6
	φ Different	I Same			7	8
		I Different			9	10

In addition, each of these categories may in principle differ with respect to logical form, in particular in the thematic relations expressed by the sentences. We will limit ourselves here to the case in which thematic relations account for the only difference between the meanings of two sentences, e.g. *Volevo farlo suonare* ("I wanted him to play" or "I wanted to have it played"), where *lo* is either the subject or the object of the verb *suonare* "to play". We will thus add an eleventh category, 1′, consisting of sentences with identical syntactic and prosodic structures but with different thematic relations.

A comparison of the predictions made by proposals such as those of Lehiste [1973] and Cooper and Paccia-Cooper [1980], in which the syntactic structure determines potential disambiguation, and a proposal in which the prosodic structure determines potential disambiguation, reveals both similarities and differences. Specifically, the syntactic and prosodic proposals make different predictions when syntactic constituency is different but prosodic constituency is constant across the two readings. The predictions of each proposal are listed in columns 5 and 6 of Table 10.1 (where a plus sign indicates that disambiguation is predicted, and a minus sign that it is not).

With respect to our additional category of sentences in which the two readings have identical syntactic and prosodic structures but different thematic relations, an approach such as that taken by Lehiste and by Cooper and Paccia-Cooper does not make explicit predictions. However, it implicitly predicts that these sentences, because they have the same syntactic structure, will not be disambiguable, a prediction made explicitly by our proposal.

In addition, the prosodic constituency analysis allows us to make distinctions beyond that of simply predicting whether disambiguation is possible or not. First of all, those sentences in which both φ and I differ should be the easiest to disambiguate. Secondly, those sentences in which either φ or I differs, but not both, should be somewhat harder to disambiguate. Since φ is particularly relevant for the domain of certain ESR's, and perhaps other prosodic phenomena such as lengthening, and I is the domain of certain other ESR's as well as intonation contours, it is difficult to determine whether a difference in one of these categories affects disambiguation more than a difference in the other category. If, however, no ESR's are applicable, we hypothesize that a difference in I's will prove more effective in disambiguation than a difference in φ's, since I boundaries always correspond to φ boundaries, yielding a combination of factors that identify φ and those that identify I.

The specific hypotheses made in connection with a prosodic constituency versus a syntactic constituency approach to ambiguity resolution are evaluated on the basis of a perception test described in the next section. To recapitulate, the hypotheses investigated are:

(i) Sentences in which I and φ structure are the same in both readings are not disambiguable even when syntactic structure differs.
(ii) Disambiguation is most likely when both I and φ are different in the two readings.
(iii) In the absence of external sandhi rule application, disambiguation is more likely when I structure differs across the two readings (though φ structure is the same) than when φ structure differs and I structure remains the same.

A fourth hypothesis was constructed to evaluate the effect of a difference in logical form on prosody:

(iv) Sentences in which the two readings differ only in thematic relations are not disambiguable.

10.4.2 Perception Test

a) Material

The test consisted of 78 pairs of ambiguous sentences distributed among the different categories of ambiguity described above. One category was missing, however; we were unable to find any Italian examples of ambiguous sentences in which syntactic, φ and I constituencies were identical in both readings but syntactic labelling differed – category 2 above.

The test was divided into two parts (Test I and Test II) such that each of the two parts contained only one sentence per pair; only the combination of the two tests gave all 156 interpretations. In both I and II, for each category of sentence, half of the sentences were presented in the meaning corresponding to one of the two readings and the other half in the meaning corresponding to the other reading. The two lists of sentences were read into a Uher 2000 tape recorder by the same native speaker of northern Italian, at a normal, relatively rapid tempo. To ensure that the person who recorded the stimuli read the sentence with the desired meaning, each ambiguous sentence was immediately preceded by an unambiguous sentence with the same syntactic structure as the intended meaning. The number of syllables and the location of stresses were also kept as similar as possible. Only the target sentences were recorded. Examples are given in (10.20).

Test I
primer: *Guido le ricerche molto seriamente.*
"I guide the research (pl.) very seriously."
target: *Guido la ricerca molto seriamente.*
"I guide the research (sg.) very seriously." (10.20 a)

Test II
primer: *Guido la prepara molto seriamente.*
"Guido prepares it very seriously."
target: *Guido la ricerca molto seriamente.*
"Guido researches it very seriously." (10.20 b)

The answer sheets consisted of 78 pairs of sentences. Each member of a pair corresponded to one of the meanings of the target sentences. Two forms of the answer sheets were used, differing only in the order of presentation of the possible meanings of each sentence. This was done in order to compensate for the possible tendency on the listener's part to consistently choose either the first or the second interpretation. The sentences on the answer sheet corresponding to the examples in (10.20) are given below.

La ricerca e guidata da me. "The research is guided by me." (10.21 a)

Guido fa ricerca. "Guido does research." (10.21 b)

b) Subjects

The subjects were 20 adult native speakers of standard Italian. All had a university education and were from the north of Italy.

c) Procedure

Each subject heard either Test I or Test II, but not both; i.e. no subject heard more than one version of any sentence. The subjects were instructed to listen to each sentence they heard, think about its interpretation, and only then look at the possible interpretations on the answer sheet. The sentence was repeated, and the subjects then marked the interpretation they thought corresponded to the sentence they had heard. If both interpretations were equally possible, the subjects were asked to indicate this explicitly. They were encouraged not to think about the sentences too long but to respond as spontaneously as possible.

d) Results

The tests were scored by assigning "+" to those answers that corresponded to the intended meaning of a given stimulus sentence, and "–" to those corresponding to the meaning that was not intended. Cases in which a subject indicated that both interpretations were possible were scored "A" and were later equally divided over the "+" and "–" values for the purposes of the analysis. Table 10.1 shows the results of the perception test, along with an example of each sentence type. The percentage of answers corresponding to the intended meanings (i.e. "+" scores) is shown for each type.

As the results show, the first two categories, those in which the syntactic labels and constituents are the same, are not disambiguable. That is, the scores are quite close to the 50% level (50% = random responses). Such a result is predicted by both the syntactic and the prosodic structure proposals.

The cases in which the two proposals make different predictions, i.e. those sentences in which syntactic structures differ across the two readings but prosodic structures are the same, clearly support the prosodic structure proposal. While the scores are slightly above the 50% level, they are far from the 100% (total disambiguation) level. Hypothesis (i) is thus confirmed.

As mentioned above, the syntactic structure hypothesis does not make specific claims about the degree of disambiguability of different sentence types, while the prosodic structure hypothesis does. If we now consider the specific claims made by the latter, we see that the categories in which both φ and I differ across the two readings are, indeed, highly disambiguable, receiving at least 90% "+" scores. They are not, however, the "most disambiguable" category, since the categories in which only I, but not φ, differs have similarly high scores of disambiguation. Hypothesis (ii) is thus disconfirmed. This advantage of I differences over φ differences, on the other hand, clearly confirms hypothesis (iii). It should be noted, furthermore, that the categories in which φ but not I differs still exhibit a significant degree of disambiguation (close to 75%), halfway between random answers (50%) and complete disambiguation (100%). From these results, we can conclude that

Table 10.1. Results of perception test. Percentage of answers corresponding to intended meaning. "S" stands for "same" and "D" for "different". A plus sign indicates that disambiguation is predicted, a minus sign that it is not

	Syntactic Constituency	Syntactic Labels	φ Structure	I Structure	Prediction of syntactic proposal	Prediction of prosodic proposal	Sample sentence	Intended meaning [%]
1.	S	S	S	S	–	–	La sua influenza mi preoccupa. "His flu worries me." "His influence worries me."	56.3
1′.	S	S	S	S	–	–	Volevo farlo suonare. "I wanted to make him play." "I wanted to have it played."	48.8
3.	D	S	S	S	+	–	Ha mangiato i pasticcini con la cioccolata. "He ate the donuts with (containing) the chocolate." "He ate the donuts (together) with the chocolate."	61.7
4.	D	D	S	S	+	–	Andrea taglia la radice e Luca la pianta. "Andrea cuts the root and Luca plants it." "Andrea cuts the root and Luca (cuts) the plant."	55.6
5.	D	S	S	D	+	+	Quando Giorgio chiama suo fratello è sempre nervoso. "When G. calls his brother, he is always nervous." "When G. calls, his brother is always nervous."	91.8
6.	D	D	S	D	+	+	Quando Franco va a trovare Maria e Lisabetta/Elisabetta torna a case molto tardi. "When F. visits Maria and Lisabetta, he comes home very late." "When F. visits Maria, Elisabetta comes home very late."	98.0
7.	D	S	D	S	+	+	Ho incontrato dei mercanti di tappeti persiani. "I met some Persian merchants of rugs." "I met some merchants of Persian rugs."	73.1

8.	D	D	D	S	+	+	La vecchia legge la regola. "The old woman reads the rule." "The old law regulates it."	78.6
9.	D	S	D	D	+	+	Ti presento il famoso architetto Davide. "I'll introduce you to the famous architect Davide." "I'll introduce you to the famous architect, Davide."	89.6
10.	D	D	D	D	+	+	La/Là cantano molto forte. "They are singing it very loudly." "There they are singing very loudly."	97.5

the most important factor in the disambiguation of sentences is a difference at the I level, the factor that the three most highly disambiguable categories have in common. A difference at the φ level, when there is no difference at the I level, leads only to partially successful disambiguation.

Finally, as suggested above, logical form does not seem to have an effect on the prosody. This is seen by the fact that the score for sentences of type 1′ represents essentially random choice of interpretations; that is, where the prosodic and syntactic structures are the same, a difference at the level of logical form does not disambiguate, in confirmation of hypothesis (iv).

Returning to the main point of this section, we can conclude that prosodic constituents predict disambiguation better than syntactic constituents. In those categories where the two analyses make different predictions, namely in sentences of type 3 and type 4 (see Table 10.1) the prediction made on the basis of prosodic structure is the correct one; disambiguation depends on a difference in prosodic structure, not syntactic structure. Furthermore, the finer distinctions explainable by prosodic structure also turn out to be important, though they are not quite those originally hypothesized. Given different prosodic structures in a pair of ambiguous sentences, we can further distinguish between those sentences in which the intonational phrases (I's) differ, independently of the phonological phrases (φ's), and those in which the I's do not differ but the φ's do; the former are more disambiguable than the latter. What was not found was a difference, within the groups with different I's, between those sentences in which the φ's were also different and those in which they were the same. Finally, the fact that hypothesis (iv) was confirmed shows that it is only the phonological (prosodic) structure that accounts for disambiguation and that logical form, and in particular thematic relations, do not have a direct effect on the disambiguation of sentences.

10.5 Conclusions

We have presented a set of rules for constructing (and restructuring) three prosodic categories above the word level: the phonological phrase, the intonational phrase and the utterance. While these rules make use of syntactic information, the resulting prosodic constituents are not necessarily isomorphic to syntactic constituents. We have demonstrated that in several cases where prosodic and syntactic structure differ, the prosodic structure makes correct predictions, specifically with regard to the positions of certain intonation breaks, while the syntactic structure fails to do so. Furthermore, it has been demonstrated that the possibility of distinguishing the meanings of ambiguous sentences depends on prosodic structure rather than on syntactic structure and thematic relations.

11. Comparative Notes on Terms and Topics in the Contributions

In an attempt to provide the reader with a way of resolving the many terminological differences in the contributions collected in this volume, we have prepared the following notes on five specific areas of terminological disagreement: Accents, Domains, Lines, Grids and Tones. The first two topics are discussed at considerably greater length than the other three; this difference reflects the relative importance of the topics – and the relative potential for confusion – both in the field of prosodic research and in the present collection. In some of the notes we outline theoretical differences which underlie choices of terminology. As much as possible, we have referred to usages and discussions in the contributions in this volume.

11.1 Accents

Stress and *accent* have long been near-synonyms for prosodic features which render some syllables acoustically more prominent than others. *Accent* is generally the broader of the two terms. Until about thirty years ago *stress* was widely used to refer to prominence realised by greater intensity, and a distinction was sometimes drawn between *stress accent* (as in English) and *pitch accent* (as in Swedish). *Stress* was assumed to be clearly distinct from the pitch phenomena involved in intonation, despite certain obvious interactions (see below, *sentence accent*); the relation between *pitch accent* and intonation was (and remains) a matter of controversy.

Significant advances in phonetic research in the 1950s (led by Fry [1955]) showed clearly the interdependence of pitch and intensity, and duration as well, in the perception of prominent syllables, and made a redefinition of *stress* necessary. Since that time, work on *stress* has taken two main directions. One is based on Bolinger's [1958] theory of pitch accent in English, which distinguished three phenomena: (*word*) *stress* (abstract prominence in lexical representation); (*pitch*) *accent* (pitch movement at stressed syllables in actual utterances, signalling pragmatic highlighting of the accented word as well as other pragmatic information); and *intonation* (other sentence-level pitch features, such as pitch direction at boundaries and the relative height of accent peaks). The second approach maintains the American structuralist insistence on distinguishing *stress* and *intonation,* abstracting away from pitch not only in lexical representation, but also in the prosodic representation of utterances [Liberman and Prince, 1977]. This view emphasises the importance of metrical or rhythmic organisation (see Sect. 11.2 "Domains"), and assumes that at the utterance level there is relative prominence, rather than either the presence or absence of accent.

The term (*pitch*) *accent* is thus applied to languages like English rather more now than it was 30 years ago. Those who define prominence in actual utterances primarily in terms of pitch movement tend to reserve the term *stress* – or *word stress,* or *lexical stress* – for the lexical abstraction, and to use the term *accent* (see the papers by Gårding, Thorsen and Cutler), or *phonological prominence* (Brown) for actual utterance prominences marked by pitch movements. (However, for *lexical stress* Gårding uses *word accent,* and Vaissière and Buxton use *stress* in all contexts. See below under "Domains" for Thorsen's use of the term *stress group*.) Those who presuppose a more abstract definition of prominence may still use the term (*pitch*) *accent* to refer to the pitch movements at stressed syllables in actual utterances (e.g. Ladd), but they conceive of *accent* (so defined) as part of the intonational structure, not as the realisation of stress. In this sense the term shares much with the traditional British term (*nuclear*) *tone* (see Sect. 11.5, "Tones").

The unresolved issue in this terminological tangle is the relation between prominence and intonation. It is now generally acknowledged that *word stress* should be considered an abstraction, but the need for an abstract representation of "metrical structure" or "relative strength" at the utterance level – i.e. a definition of prominence in actual utterances that does not make reference to pitch features – is disputed.

The terms *sentence accent, sentence stress, nuclear stress, nuclear accent, nucleus, focal accent,* and *tonic* all apply to a phenomenon at the conceptual boundary between prominence and intonation; Thorsen discusses some of the correspondences. Those who base their definition of *accent* on pitch movement sometimes reject the concept of *sentence accent,* arguing that there is no basis for designating one accent in a sentence as more prominent than the rest (e.g. Brown); or they may acknowledge a separate phonetic basis for the distinction (e.g. Gårding) or suggest that languages or utterances vary in this respect (e.g. Thorsen). Those who adopt a more abstract basis for utterance prominence, by contrast, generally assume the existence of one most prominent syllable in a given domain (see "Domains"). Insofar as that domain need not be a sentence, *nuclear stress* or *nuclear accent* is probably a more appropriate term than *sentence stress* or *sentence accent.*

The functions of *sentence accent* (or just *accent,* in descriptions which do not recognise distinctions among accents) are often described with such terms as *highlighting, focus, contrast, new information, comment,* etc. Conceptions of these vary very widely, and interact with the definitions of *accent* discussed above; one central issue is whether prominence involves primarily hierarchical or primarily linear structure (see also "Domains"). A linear model favours the assumption that a string of equal *accents* is assigned to individual words to highlight them (e.g. Brown); obvious phonetic inequalities between accents are attributed to other factors such as *contrast* or *emphasis* (Thorsen's term is *emphasis for contrast*). In a hierarchical model, on the other hand, degrees of relative prominence are specified within prosodic domains, and focus or highlighting can thus apply to domains larger than the word.

Deaccenting (the term is used by Cutler) is the other side of *focus*, and has two corresponding definitions: in a more linear model it is absence of *accent*, i.e. absence of pitch prominence; in a hierarchical model it refers to the reduction of prominence within the prosodic structure. Either way it signals that a word or constituent is *given, old information, presupposed, contextually inferable, topic,* etc. – see Brown for discussion.

11.2 Domains

One of the most important characteristics of prosody is that it involves organisation above the segment, or (viewed differently) phenomena that extend longer in time than a single segment. In analysing these phenomena, it is common to make use of the notion of *domains* of various sizes, i.e. chunks of structure over which prosodic phenomena may be defined. However, different investigators have adopted such different approaches to the data that it is extremely difficult to compare terminology for domains of different sizes.

The major disagreements concern the nature of the hierarchical structure in prosody, and the extent to which prosodic domains can be defined in terms of concrete "linear" features such as pauses. The scheme formalised by Halliday [1967 b], which assumes a relatively "shallow" hierarchy, organises syllables into *feet* and feet into *tone groups*. Halliday took over the term *foot* from Abercrombie [1964]; a *foot* consists of a stressed syllable (actually a "salient" syllable in Halliday's terminology) followed optionally by one or more unstressed syllables, which may be full or reduced. A *tone group* consists of one or more feet, one of which (often the last) is the location of the *tonic* (which corresponds to *sentence accent* in other accounts; see "Accents"). An important characteristic of Halliday's system is that it is intended to describe utterances, and hence assigns prosodic structure not as an abstract property of words and sentences, but as a dimension of their spoken realisation. Within this system it is therefore possible for a particular word or sentence to be assigned different structures in different utterances.

The term *foot* is particularly likely to cause confusion. Hirst and Nespor and Vogel use it in the sense developed by Selkirk [1978 b, 1980]; this usage is based on a "deeper" hierarchical structure than Halliday's (Hirst and Nespor and Vogel discuss this structure in detail). Selkirk's *foot* (in fact, Selkirk originally used the term *stress foot* specifically in order to distinguish her usage from that of Abercrombie and Halliday) was in turn based on the *foot* which figured in Liberman and Prince's [1977] hierarchical model (see Sect. 11.3, "Grids"). Liberman and Prince used the term to describe a branching structure comprising one S (strong) syllable followed optionally by one or more W (weak) syllables. In Selkirk's system it is roughly equivalent to any non-reduced syllable with optionally one or more reduced syllables following it. Thus in both these systems the term describes a smaller unit than the Abercrombie/Halliday *foot.* In fact, the latter corresponds more closely to Selkirk's *prosodic word* (Vaissière also uses this term; Nespor and Vogel's version of Selkirk's terminology, however, refers to it as a *phonological word;* cf. also Thorsen's *stress group*). *Prosodic word* was also adopted by Selkirk from Liberman and Prince's terminology (though they referred to it also as *mot*); it diverges from the Abercrombie/Halliday *foot* only in the case of words with secondary stress (e.g. "photogenic"), which are described as two feet by Abercrombie or Halliday, and one prosodic word by Selkirk. The number of feet which Selkirk's system assigns to a prosodic word depends on the number of unreduced syllables; thus a word like "reconciliation" is described as one prosodic word of three two-syllable feet in Selkirk's system, while in an utterance it would most likely be spoken with two feet (the first of four, the second of two syllables) according to an Abercrombie/Halliday description.

The term *tone group* is used by Brown (although she rejects Halliday's definition of *tonic* based on *tone groups;* see "Accents"). In her use it seems to correspond roughly to Hirst's or Gårding's *phrase,* though again it is clear that there are considerable theoretical differences between these authors. How either term corresponds to Nespor and Vogel's (or Selkirk's) more extensive and more formally defined terminology is not completely clear. Both *phrase* and *tone group* might be applied to domains which Nespor and Vogel would distinguish as *phonological phrase* and *intonational phrase* respectively, though in general it seems correct to say that *intonational phrase* corresponds to *tone group,* while *phonological phrase* is intermediate in size between *tone group* and (any definition of) *foot.*

Tone group and (*intonational*) *phrase* define domains in terms of structural characteristics of the prosody itself. A different basis for defining comparably sized chunks is reflected in the term *breath group,* first made current by Lieberman [1967] and used in this volume by Vaissière. It is often assumed (as is explicitly stated by Vaissière) that there is a relation between *breath groups* and *declination* (see Sect. 11.4, "Lines"). While the use of the term *breath group* need not preclude a structural as opposed to an acoustic/physiological definition, it tends to be used in the context of relatively "concrete" models of prosody which place little or no emphasis on hierarchical structure.

11.3 Grids

The main use of the term *grid* in recent prosodic work is in connection with rhythmic organisation; the Liberman and Prince [1977] *metrical grid* is an attempt to formalise the phenomenon of *stress timing* (sometimes simply referred to as *isochrony*), the tendency apparent in certain languages for equally prominent syllables to occur at regular intervals. *Stress timing* is usually contrasted with *syllable timing,* the apparent tendency in certain other languages for syllables to occur at regular intervals. Buxton discusses the question of *stress timing* in English.

In Gårding's contribution the term *tonal grid* is used to designate the assemblage of intonational reference lines (see Sect. 11.4, "Lines") used for scaling the peaks and valleys of utterance contours.

11.4 Lines

Reference is made in several of the contributions in this volume to *topline, baseline, reference line, zero line, focus line, declination line,* etc. Most of these are similar in that they are abstractions which simplify the complexity of actual utterance contours by indicating an overall direction or shape. Illustrative examples are found in the contributions by Gårding, Thorsen and Vaissière.

Declination lines are a special case, since the declination tendency is often supposed to be universal and related to syntactic structure (see Vaissière's discussion). *Resetting the baseline* (Vaissière) refers to the raising of pitch after a major syntactic break and the beginning of a new downward trend. *Declination* is usually distinguished from *downstep;* the latter refers to a more "local" phenomenon than the former, namely the lowering of one high tone relative to a preceding high tone (see Sect. 11.5 and Ladd's discussion of this).

11.5 Tones

Tone is among the most ambiguous words in the prosodist's lexicon. Important distinct meanings are (a) *lexical tone:* distinctive pitch features that identify morphemes in tone languages such as Chinese or Igbo; (b) *nuclear tone:* the distinctive pitch features associated with the main accented syllable of a phrase or utterance in an "intonation language" such as English (see

"Accents"); (c) *tones:* abstract elements in terms of which pitch features may be specified; e.g. abstract highs and lows specifying rises and falls. The first meaning is mentioned in Hirst's and in Nespor and Vogel's contributions. Ladd discusses the second meaning, but uses the term *nuclear accent* rather than *nuclear tone.* Only the third meaning figures prominently in the discussions in the present volume. Ladd and Hirst both argue for models of intonation that use abstract *tones;* so does Gårding, although she does not use the term *tone,* but *tonal grid* (see "Grids"). Thorsen argues that a model incorporating abstract *tones* does not adequately represent Danish. (For the term *tone group* see "Domains".)

References

Abercrombie, D. (1964): "Syllable Quantity and Enclitics in English", in *In Honour of Daniel Jones,* ed. by D. Abercrombie, D. B. Fry, P. A. D. MacCarthy, N. C. Scott, J. L. M. Trim (Longmans, London)

Allen, G. D. (1972): The location of rhythmic stress beats in English: An experimental study; I and II. Lang. Speech **15**, 72–100 and 179–195

Baars, B. J., Motley, M. T., MacKay, D. G. (1975): Output editing for lexical status in artificially elicited slips of the tongue. J. Verb. Learn. Verb. Behav. **14**, 382–391

Basbøll, H. (1978): "A Note on Boundaries and Stress Rules in Danish Phonology", in *Nordic Prosody,* ed. by E. Gårding, G. Bruce, R. Bannert (Gleerup, Lund) pp. 65–71

Black, J. W., Tosi, O., Singh, S., Takefuta, Y. (1966): A study of pauses in oral reading of one's native language and in English. Lang. Speech **9**, 237–241

Blumstein, S., Cooper, W. E. (1974): Hemispheric processing of intonation contours. Cortex **10**, 148–158

Bolinger, D. L. (1958): A theory of pitch accent in English. Word **14**, 109–149

Bolinger, D. L. (1965): "Pitch Accent and Sentence Rhythm", in *Forms of English: Accent, Morpheme, Order* (Harvard University Press, Cambridge, MA) pp. 139–180

Bolinger, D. L. (1970): "Relative Height", in *Intonation,* ed. by D. Bolinger (Penguin, Harmondsworth) pp. 137–153

Bolinger, D. L. (1972): Accent is predictable (if you're a mind-reader). Language **48**, 633–644

Bolinger, D. L. (1982): Intonation and its parts. Language **58**, 505–533

Botinis, A. (1982): Stress in modern Greek. Work. Pap. Linguist., Univ. Lund **22**, 27–38

Brazil, D., Coulthard, M., Johns, C. (1980): *Discourse Intonation and Language Teaching* (Longmans, London)

Brown, G., Currie, K., Yule, G. (1979): "The Intonation of Scottish English", final report on SSRC Project HR 3601

Brown, G., Currie, K., Kenworthy, J. (1980): *Questions of Intonation* (Croom Helm, London)

Brown, K. (1980): "Grammatical Incoherence", in *Temporal Variables in Speech: Studies in Honour of Frieda Goldman-Eisler,* ed. by H. W. Dechert, M. Raupach (Mouton, The Hague) pp. 28–37

Bruce, G. (1977): *Swedish Word Accents in Sentence Perspective,* Travaux de l'Institut de Linguistique de Lund, 12 (Gleerup, Lund)

Bruce, G. (1981): Om prosodisk koppling i svenska. Umeå Stud. Humanities **37**, 64–78

Bruce, G. (1982): Developing the Swedish intonation model. Work. Pap. Linguist., Univ. Lund **22**, 51–114

Bruce, G., Gårding E. (1978): "A Prosodic Typology for Swedish Dialects", in *Nordic Prosody,* ed. by E. Gårding, G. Bruce, R. Bannert (Gleerup, Lund) pp. 219–228

Burden, R. L., Faires, J. D., Reynolds, A. C. (1981): *Numerical Analysis,* 2nd ed. (Prindle, Weber & Schmidt, Boston)

Butterworth, B. (1980): "Evidence from Pauses in Speech", in *Language Production I: Speech and Talk,* ed. by B. Butterworth (Academic, London) pp. 155–176

Buxton, H. (unpublished): "Rhythm and Stress in Speech", Ph. D Thesis, University of Cambridge

Carlson, R., Granström, B. (1973): "Word Accent, Emphatic Stress, and Syntax in a Synthesis by Rules Scheme for Swedish", Speech Transmission Laboratory (Stockholm), Quarterly Progress and Status Report, 2/3–1973, pp. 31–35

Carlson, R., Erikson, Y., Granström, B., Lindblom, B., Rapp, K. (1975): "Neutral and Emphatic Stress Patterns in Swedish", in *Speech Communication, Vol. 2: Speech Production and Synthesis by Rules,* ed. by G. Fant (Almqvist & Wiksell, Stockholm) pp. 209–218

Chafe, W. L. (1970): *Meaning and the Structure of Language* (Chicago University Press, Chicago)

Chafe, W. L. (1974): Language and consciousness. Language, **50**, 111–133

Chafe, W. L. (1976): "Givenness, Contrastiveness, Definiteness, Subjects, Topics and Points of view", in *Subject and Topic,* ed. by C. Li (Academic, New York) pp. 25–55

Chafe, W. L. (1979): "The Flow of Thought and the Flow of Language", in *Syntax and Semantics 12: Discourse and Syntax,* ed. by T. Givon (Academic, New York)

Chen, M. (1970): Vowel length variation as a function of the voicing of the consonant environment. Phonetica **22**, 129–159

Chomsky, N. (1955): *The Logical Structure of Linguistic Theory* (Plenum, New York) [revised 1973]

Chomsky, N. (1965): *Aspects of the Theory of Syntax* (MIT Press, Cambridge, MA)

Chomsky, N. (1981): *Lectures on Government and Binding* (Foris, Dordrecht)

Chomsky, N., Halle, M. (1968): *The Sound Pattern of English* (Harper & Row, New York)

Clark, H. H., Clark, E. V. (1977): *Psychology and Language* (Harcourt Brace Jovanovich, New York)

Clark, H. H., Haviland, S. E. (1977): "Comprehension and the Given-New Contract", in *Discourse Production and Comprehension,* ed. by R. O. Freedle (Ablex, Norwood, NJ)

Classe, A. (1939): *The Rhythm of English Prose* (Blackwell, Oxford)

Clements, G. N. (1975): *"Tone as Speech Melody",* unpublished manuscript, Massachusetts Institute of Technology

Clements, G. N. (1976): *"Vowel-Harmony in Non-linear Generative Phonology: An Autosegmental Approach",* Indiana University Linguistics Club

Clemmer, E. J., O'Connell, D. C., Loui, W. (1979): Rhetorical pauses in oral reading. Lang. Speech **22**, 397–405

Cohen, A., 't Hart, J. (1967): On the anatomy of intonation. Lingua **19**, 177–192

Coker, C. H., Umeda, N. (1973): Automatic synthesis from ordinary English text. IEEE Trans. Audio Electroacoust. **AU-21**, 293–297

Collier, R. (1975 a): Perceptual and linguistic tolerance in intonation. Int. Rev. Appl. Linguist. **13**, 293–308

Collier, R. (1975 b): Physiological correlates of intonation patterns. J. Acoust. Soc. Am. **58**, 249–255

Collier, R., 't Hart, J. (1975): "The Role of Intonation in Speech Perception", in *Structure and Process in Speech Perception,* ed. by A. Cohen, S. G. Nooteboom (Springer, Berlin, Heidelberg, New York) pp. 107–121

Contini, M., Boë, L. J. (1975): "Etude Quantitative de l'Intonation en Français", in Proc. 8th Int. Congr. Phonetic Sciences, Leeds

Cooper, W. E. (1976): "Syntactic Control of Timing in Speech Production", Ph. D. Thesis, Massachusetts Institute of Technology

Cooper, W. E., Paccia-Cooper, J. (1980): *Syntax and Speech* (Harvard University Press, Cambridge, MA)

Cooper, W. E., Sorensen, J. M. (1981): *Fundamental Frequency in Sentence Production* (Springer, Berlin, Heidelberg, New York)

Cowan, J. M., Bloch, B. (1948): An experimental study of pause in English grammar. Am. Speech **23**, 89–99

Cruttenden, A. (1981): Falls and rises: Meanings and universals. J. Linguist. **17**, 77–91

Crystal, D. (1969): *Prosodic Systems and Intonation in English* (Cambridge University Press, Cambridge)

Crystal, D. (1975): *The English Tone of Voice* (Edward Arnold, London)

Currie, K. L. (1980): An initial search for tonics. Lang. Speech **23**, 329–350
Currie, K. L. (1981): Further experiments in the "search for tonics". Lang. Speech **24**, 1–28
Cutler, A. (1976): Phoneme-monitoring reaction time as a function of preceding intonation contour. Percept. Psychophys. **20**, 55–60
Cutler, A. (1980 a): "Syllable Omission Errors and Isochrony", in *Temporal Variables in Speech: Studies in Honour of Frieda Goldman-Eisler,* ed. by. H. W. Dechert, M. Raupach (Mouton, The Hague) pp. 183–190
Cutler, A. (1980 b): "Errors of Stress and Intonation", in *Errors in Linguistic Performance,* ed. by V. A. Fromkin (Academic, New York) pp. 67–80
Cutler, A., Clifton, C. E. (1983): "The Use of Prosodic Information in Word Recognition", in *Attention and Performance X,* ed. by H. Bouma, D. G. Bouwhuis (Erlbaum, Hillsdale, NJ)
Cutler, A., Fodor, J. A. (1979): Semantic focus and sentence comprehension. Cognition **7**, 49–59
Cutler, A., Foss, D. J. (1977): On the role of sentence stress in sentence processing. Lang. Speech **20**, 1–10
Dahl, O. (1976): "What is New Information? in *Reports on Text-Linguistics: Approaches to Word Order,* ed. by N. E. Enkvist, V. Kohonen (Text Linguistics Research Group, Åbo)
Dascălu, L. (1979): On the intonation of questions in Romanian: The rising pattern. Rev. Roum. Linguist. **24**, 35–41
Deese, J. (1980): "Pauses, Prosody and the Demands of Production in Language", in *Temporal Variables in Speech: Studies in Honour of Frieda Goldman-Eisler,* ed. by H. W. Dechert, M. Raupach (Mouton, The Hague) pp. 69–84
Delattre, P. (1962): Some factors of vowel duration and their cross-linguistic validity. J. Acoust. Soc. Am. **34**, 1141–1143
Delattre, P. (1963): Comparing the prosodic features in English, German, Spanish and French. Int. Rev. Appl. Linguist. **1**, 193–210
Delattre, P. (1966): A comparison of syllable length conditioning among languages. Int. Rev. Appl. Linguist. **4**, 183–198
Dell, G. S., Reich, P. A. (1981): Stages in sentence production: An analysis of speech error data. J. Verb. Learn. Verb. Behav. **20**, 611–629
de Pijper, J. R. (1979): Close-copy stylizations of British English intonation contours. IPO Annu. Prog. Rep. **14**, 66–71
de Saussure, F. (1916): *Cours de Linguistique Générale* (Payot, Paris)
Dogil, G. (1981): "Elementary Accent Systems", in *Phonologica 1980,* ed. by W. U. Dressler, O. Pfeiffer, J. Rennison (Innsbrücker Beiträge zur Sprachwissenschaft, Innsbruck) pp. 89–100
Donovan, A. (1977): "Perceptual Isochrony", Conference of the Speech Group of the Institute of Acoustics (U.K.), University of Sussex, July
Donovan, A., Darwin, C. J. (1979): "The Perceived Rhythm of speech", Proc. 9th Int. Congr. Phonetic Sciences, Copenhagen Vol. 2, pp. 268–274
Duez, D. (1982): Silent and non-silent pauses in three speech styles. Lang. Speech **25**, 11–28
Fairbanks, G., Hoaglin, L. W. (1941): An experimental study of the durational characteristics of the voice during the expression of emotion. Speech Monogr. **8**, 85–90
Faure, G., Hirst, D., Chafcouloff, M. (1980): "Rhythm in English: Isochronism, Pitch and Perceived Stress", in *The Melody of Language,* ed. by L. Waugh, C. van Schooneveld (University Park Press, Baltimore) pp. 71–79
Fay, D. A. (1980): "Transformational Errors", in *Errors in Linguistic Performance,* ed. by V. A. Fromkin (Academic, New York) pp. 111–122
Fodor, J. A., Bever, T. G., Garrett, M. F. (1974): *The Psychology of Language* (McGraw-Hill, New York)
Ford, K. (1975): "Downstep in Kikuyu", unpublished manuscript
Foss, D. J. (1969): Decision processes during sentence comprehension: Effects of lexical item difficulty and position upon decision times. J. Verb. Learn. Verb. Behav. **8**, 547–562
Fowler, C. (1977): "Timing Control in Speech Production", Ph. D. Thesis, University of Connecticut (Distributed by Indiana University Linguistics Club)

Fowler, C. (1979): "Perceptual centers" in speech production and perception. Percept. Psychophys. **25**, 375–388

Frazier, L., Fodor, J. D. (1978): The sausage machine: A new two-stage parsing model. Cognition **6**, 291–325

Fromkin, V. A. (1972): Tone features and tone rules. Stud. Afr. Linguist. **3**, 47–76

Fromkin, V. A. (1977): "Putting the EmPHAsis on the Wrong SylLABle", in *Studies in Stress and Accent,* ed. by L. M. Hyman (University of Southern California, Los Angeles) pp. 15–26

Fry, D. B. (1955): Duration and intensity as physical correlates of linguistic stress. J. Acoust. Soc. Am. **27**, 765–769

Fujisaki, H., Sudo, H. (1971): A generative model for the prosody of connected speech in Japanese. Annu. Rep. Eng. Res. Inst., Univ. Tokyo **30**, 75–80

Fujisaki, H., Sugito, M. (1976): Temporal organisation of articulatory and phonatory controls in the realization of Japanese word accent. Annu. Bull. Res. Inst. Logopedics Phoniatrics, Univ. Tokyo **10**, 176–189

Fujisaki, M., Hirose, K., Ohta, K. (1979): Acoustic features of the fundamental frequency contours of declarative sentences. Annu. Bull. Res. Inst. Logopedics Phoniatrics, Univ. Tokyo **13**, 163–172

Garde, P. (1968): *L'Accent* (Presses Universitaires de France, Paris)

Gårding, E. (1962): Olikheter mellan svensk och engelsk intonation i påståendesatser. Mod. Språk **56**, 269–280

Gårding, E. (1975a): "Toward a Prosodic Typology for Swedish Dialects", in *The Nordic Languages and Modern Linguistics,* Vol. 2, ed. by K.-H. Dahlstedt (Almqvist & Wiksell, Stockholm) pp. 466–474

Gårding, E. (1975b): The influence of tempo on rhythmic and tonal patterns in three Swedish dialects. Work. Pap. Linguist., Univ. Lund **12**, 71–83

Gårding, E. (1977): "The Importance of Turning Points for the Pitch Patterns of Swedish Accents", in *Studies in Stress and Accent,* ed. by L. M. Hyman (University of Southern California, Los Angeles) pp. 27–35

Gårding, E. (1979): Sentence intonation in Swedish. Phonetica **36**, 207–215

Gårding, E. (1980): Moderator's summary, Symp. 7: "The relation between sentence prosody and word prosody", Proc. 9th Int. Congr. Phonetic Sciences, Copenhagen, Vol. 3, pp. 283–287

Gårding, E. (1981): Contrastive prosody: A model and its application. Stud. Linguist. **35**, 146–165

Gårding, E. (1982): Prosodic expressions and pragmatic categories. Work. Pap. Linguist., Univ. Lund **22**, 117–136

Gårding, E., Lindblad, P. (1973): Constancy and variation in Swedish word accent patterns. Work. Pap. Linguist., Univ. Lund **7**, 36–110

Gårding, E., Botinis, A., Touati, P. (1982): A comparative study of Swedish, Greek and French intonation. Work. Pap. Linguist., Univ. Lund **22**, 137–152

Garrett, M. F. (1975): "The Analysis of Sentence Production", in *Psychology of Learning and Motivation,* Vol. 9, ed. by G. Bower (Academic, New York) pp. 133–177

Gay, T. (1978): Physiological and acoustic correlates of perceptual stress. Lang. Speech **21**, 347–353

Goffman, E. (1981): "Radio Talk", in *Forms of Talk* (Blackwell, Oxford), pp. 197–327

Goldman-Eisler, F. (1968): *Psycholinguistics: Experiments in Spontaneous Speech* (Academic, London)

Goldman-Eisler, F. (1972): Pauses, clauses, sentences. Lang. Speech **15**, 103–113

Goldsmith, J. (1976): "Autosegmental Phonology", Ph. D. Thesis, Massachusetts Institute of Technology (Distributed by Indiana University Linguistics Club)

Grosjean, F., Collins, M. (1979): Breathing, pausing and reading. Phonetica **36**, 98–114

Grosjean, F., Deschamps, A. (1973): Analyse des variables temporelles du Français spontane. II. Comparaison du Français oral dans la description avec l'Anglais (description) et avec le Français (interview radiophonique) Phonetica **28**, 191–226

Grosjean, F., Deschamps, A. (1975): Analyse contrastive des variables temporelles de l'Anglais et du Français: Vitesse de parole et variables composantes, phénomène d'hesitation. Phonetica **31**, 144–413

Grundstrom, A., Léon, P. (1973): *Interrogation et Intonation,* Studia Phonetica, 8 (Didier, Paris)

Hadding-Koch, K. (1961): *Acoustico-Phonetic Studies in the Intonation of Southern Swedish* (Gleerup, Lund)

Halle, M., Vergnaud, J. R. (1982): "Foundations of Generative Phonology", Phonology Seminar organised by GLOW, Paris

Halliday, M. A. K. (1963): The tones of English. Arch. Linguist. **15**, 1–28

Halliday, M. A. K. (1967 a): Notes on transitivity and theme in English, Part 2. J. Linguist. **3**, 199–244

Halliday, M. A. K. (1967 b): *Intonation and Grammar in British English* (Mouton, The Hague)

Halliday, M. A. K. (1970): *A Course in Spoken English: Intonation* (Oxford University Press, Oxford)

Han, M. S. (1961): "Japanese Phonology: An Analysis Based upon Sound Spectrograms", Ph. D. Thesis, University of Texas at Austin

Haugen, E., Joos, M. (1952): Tone and intonation in East Norwegian. Acta Philol. Scand. **22**, 41–64; also in D. L. Bolinger (ed.): *Intonation* (Penguin, Harmondsworth) pp. 414–436

Haviland, S. E., Clark, H. H. (1974): What's new? Acquiring new information as a process in comprehension. J. Verb. Learn. Verb. Behav. **13**, 512–520

Hayes, B. (1979): "Extrametricality", in Papers on Syllable Structure, Metrical Structure and Harmony Processes, MIT Working Papers on Linguistics, Vol. 1, ed. by K. Safir

Hayes, B. (1981): "A Metrical Theory of Stress Rules", Ph. D. Thesis, Massachusetts Institute of Technology (Distributed by Indiana University Linguistics Club)

Heger, S. (1981): *Sprog og Lyd – Elementaer Dansk Fonetik* (Akademisk, Copenhagen)

Hill, D. R., Jassem, W., Witten, I. H. (1978): "A Statistical Approach to the Problem of Isochrony in Spoken British English", Man-Machine Systems Laboratory Report 78/27/6, University of Calgary

Hill, D. R., Witten, I. H., Jassem, W. (1978): "Some Results from a Preliminary Study of British English Speech Rhythms", Man-Machine Systems Laboratory Report 78/26/5, University of Calgary

Hirose, K., Fujisaki, H., Sugito, M. (1978): Word accent in Japanese and English: A comparative study of acoustic characteristics in disyllabic words. Annu. Bull. Res. Inst. Logopedics Phoniatrics, Univ. Tokyo **12**, 141–148

Hirst, D. J. (1977): *Intonative Features: A Syntactic Approach to English Intonation* (Mouton, The Hague)

Hirst, D. J. (1979): Pitch features for tone intonation. Trav. Inst. Phon. Aix **6**, 177–191

Hirst, D. J. (1981 a): Un modèle de production de l'intonation. Trav. Inst. Phon. Aix **7**, 297–311

Hirst, D. J. (1981 b): "Prosodie et Structures de Données en Phonologie", in *Représentations Formelles en Phonologie,* ed. by F. Dell, D. J. Hirst, J.-R. Vergnaud (Hermann, Paris)

Hirst, D. J. (1981 c): "Phonological Implications of a Production Model of Intonation", in *Phonologica 1980,* ed. by W. U. Dressler, O. Pfeiffer, J. Rennison (Innsbrücker Beiträge zur Sprachwissenschaft, Innsbruck) pp. 195–202

Hirst, D. J. (in press): "Phonological and Acoustic Parameters of English Intonation", in *Intonation in Discourse,* ed. by C. Johns-Lewis (Croom Helm, London)

Hixon, T. J., Klatt, D. H., Mead, J. (1971): Influence of forced transglottal pressure change on voice fundamental frequency. J. Acoust. Soc. Am. **49**, 105

Ho, A. T. (1977): Intonation variation in a Mandarin sentence of three expressions: Interrogative, exclamatory and declarative. Phonetica **34**, 446–457

Hombert, J. M. (1974): Universals of downdrift: Their basis and significance for a theory of tone. Stud. Afr. Linguist. **5**, 169–183

Hooper, J. B. (1972): The syllable in phonological theory. Language **48**, 525–540

House, A. S., Fairbanks, G. (1953): The influence of consonant environment upon the secondary acoustical characteristics of vowels. J. Acoust. Soc. Am. **25**, 105–113

Isačenko, A., Schädlich, H.-J. (1970): *A Model of Standard German Intonation* (Mouton, The Hague)
Jespersen, O. (1933): *Lehrbuch der Phonetik* (Teubner, Leipzig)
Jones, D. (1960): *An Outline of English Phonetics,* 9th ed. (Heffers, Cambridge)
Kahn, D. (1976): "Syllable-Based Generalisations in Phonology", Ph. D. Thesis, Massachusetts Institute of Technology (Distributed by Indiana University Linguistics Club)
Kingdon, R. (1958): *The Groundwork of English Intonation* (Longmans, London)
Kiparsky, P. (1979): Metrical structure assignment is cyclic. Linguist. Inq. **10**, 421–441
Klatt, D. H. (1979): "Synthesis by Rule of Segmental Durations in English Sentences", in *Frontiers of Speech Communication Research,* ed. by B. Lindblom, S. Öhman (Academic, London)
Klein, W. (1980): Der Stand der Forschung zur deutschen Satzintonation. Linguist. Ber. **68**, 3–33
Kloker, D. (1975): Vowel and sonorant lengthening as cues to phonological phrase boundaries. J. Acoust. Soc. Am. **57**, 533
Kozhevnikov, V. A., Chistovich, L. A. (1965): *Speech: Articulation and Perception* (U.S. Dept. of Commerce, Washington, DC) JPRS, 30, 543
Ladd, D. R. (1980): *The Structure of Intonational Meaning: Evidence from English* (Indiana University Press, Bloomington)
Ladd, D. R. (1981): "On Intonational Universals", in *The Cognitive Representation of Speech,* ed. by T. Myers, J. Laver, J. Anderson (North-Holland, Amsterdam) pp. 389–397
Ladd, D. R. (in press): Phonological features of intonational peaks. Language
Langendoen, D. T. (1975): Finite state parsing of phrase structure languages and the status of readjustment rules in the grammar. Linguist. Inq. **6**, 533–554
Laver, J. D. M. (1973): "The Detection and Correction of Slips of the Tongue", in *Speech Errors as Linguistic Evidence,* ed. by V. A. Fromkin (Mouton, The Hague) pp. 132–143
Lehiste, I. (1960): An acoustic-phonetic study of internal open juncture. Phonetica **5**, Suppl. 1–54
Lehiste, I. (1965): The function of quantity in Finnish and Estonian. Language **41**, 447–456
Lehiste, I. (1970): *Suprasegmentals* (MIT Press, Cambridge, MA)
Lehiste, I. (1972): The timing of utterances and linguistic boundaries. J. Acoust. Soc. Am. **51**, 2018–2024
Lehiste, I. (1973): Phonetic disambiguation of syntactic ambiguity. Glossa **7**, 107–122
Lehiste, I. (1975): "The Phonetic Structure of Paragraphs", in *Structure and Process in Speech Perception,* ed. by A. Cohen, S. G. Nooteboom (Springer, Berlin, Heidelberg, New York) pp. 195–203
Lehiste, I. (1977): Isochrony reconsidered. J. Phon. **5**, 253–263
Levelt, W. J. M. (1983): Monitoring and self-repair in speech. Cognition
Liberman, M. Y. (1975): "On the Intonational System of English", Ph. D. Thesis, Massachusetts Institute of Technology (Distributed by Indiana University Linguistics Club)
Liberman, M. Y., Prince, A. (1977): On stress and linguistic rhythm. Linguist. Inq. **8**, 249–336
Liberman, M. Y., Sag, I. (1974): "Prosodic Form and Discourse Function", Papers from the Tenth Regional Meeting, Chicago Linguistic Society pp. 416–426
Liberman, M. Y., Streeter, L. (1978): The use of nonsense-syllable mimicry in the study of prosodic phenomena. J. Acoust. Soc. Am. **63**, 231–233
Lieberman, P. (1963): Some effects of semantic and grammatical context on the production and perception of speech. Lang. Speech **6**, 172–187
Lieberman, P. (1967): *Intonation, Perception and Language* (MIT Press, Cambridge, MA)
Lindblom, B. (1968): "Temporal Organisation of Syllable Production", Speech Transmission Laboratory (Stockholm) Quarterly Progress and Status Report, 2–1968, pp. 1–5
Maclay, H., Osgood, C. E. (1959): Hesitation phenomena in spontaneous speech. Word **15**, 19–44
Maeda, S. (1974): "A Characterization of Fundamental Frequency Contours of Speech", Research Laboratory of Electronics (MIT), Quarterly Progress Report, **114**, pp. 193–211

Maeda, S. (1976): "A Characterization of American English Intonation", Ph. D. Thesis, Massachusetts Institute of Technology
Magno-Caldognetto, E., Ferrero, F. E., Lavagnoli, C., Vagges, K. (1978): F_0 contours of statements, yes-no questions and wh-questions of two regional varieties of Italian. J. Ital. Linguist. **3**, 57–68
Marcel, P. (1971): La durée des voyelles de l'Italien à partir de l'analyse oscillographique. Trav. Inst. Phon. Strasbourg **3**, 147–157
Marcus, S. M. (1976): "Perceptual Centres", Ph. D. Thesis, University of Cambridge
Marslen-Wilson, W. D., Welsh, A. (1978): Processing interactions and lexical access during word recognition in continuous speech. Cognit. Psychol. **10**, 29–63
Marslen-Wilson, W. D., Levy, E., Tyler, L. K. (1982): "Producing Interpretable Discourse: The Establishment and Maintenance of Reference", in *Speech, Place and Action: Studies in Deixis and Related Topics,* ed. by R. J. Jarvella, W. Klein (Wiley, Chichester) pp. 339–378
Martin, J. G. (1970): Rhythm-induced judgements of word stress in sentences. J. Verb. Learn. Verb. Behav. **9**, 627–633
Martin, J. G. (1972): Rhythmic (hierarchical) versus serial structure in speech and other behavior. Psychol. Rev. **79**, 487–509
Martin, J. G. (1979): Rhythmic and segmental perception are not independent. J. Acoust. Soc. Am. **65**, 1286–1297
Martin, J. G., Bunnell, H. T. (1981): Perception of anticipatory coarticulation effects. J. Acoust. Soc. Am. **69**, 559–567
Martin, J. G., Mills, C. B., Meltzer, R. H., Shields, J. L. (1980): Anticipatory coarticulation and reaction time to phoneme targets in spontaneous speech. Phonetica **37**, 159–168
Martin, P. (1975): Analyse phonologique de la phrase française. Linguistics **146**, 35–68
Martin, P. (1980): "Sur les Principes d'une Théorie Syntaxique de l'Intonation", in *Problèmes de Prosodie, Vol. 1: Approches Théoriques,* ed. by P. Léon, M. Rossi (Didier, Paris)
McCarthy, J. (1979): On stress and syllabification. Linguist. Inq. **10**, 443–465
McCarthy, J. (1982): Non-linear phonology: An overview. GLOW Newsl. **8**, 63–77
Meltzer, R. H., Martin, J. G., Mills, C. B., Imhoff, D. L., Zohar, D. (1976): Reaction time to temporally-displaced phoneme targets in continuous speech. J. Exp. Psychol. Hum. Percept. Perform. **2**, 277–290
Menn, L. (1976): "Pattern, Control and Contrast in Beginning Speech: A Case Study in the Development of Word Form and Word Function", Ph. D. Thesis, University of Illinois
Mitsuya, F., Sugito, M. (1978): A study of the accentual effect on segmental and moraic duration in Japanese. Annu. Bull. Res. Inst. Logopedics Phoniatrics, Univ. Tokyo **12**, 97–112
Morton, J., Marcus, S. M., Frankish, C. (1976): Perceptual centres (P-centres). Psychol. Rev. **83**, 405–408
Nakatani, L. H., O'Connor, K. D., Aston, C. H. (1981): Prosodic aspects of American English speech rhythm. Phonetica **38**, 84–106
Nespor, M. A. (1977): "Segment and Word Durational Correlates of Syntactic Boundaries in Italian", in *Explorations in Language and Linguistics,* ed. by R. Rodman (University of North Carolina, Chapel Hill, NC)
Nespor, M. A., Vogel, I. (1982): "Prosodic Domains and External Sandhi Rules", in *The Structure of Phonological Representations,* Part I, ed. by H. v. d. Hulst, N. Smith (Foris, Dordrecht) pp. 225–255
Nespor, M. A., Vogel, I. (in press): "Prosodic Hierarchy and Speech Perception", in Proceedings of the Conference on the Perception of Language, Accademia della Crusca, Florence, 1980
Newman, S. S. (1946): On the stress system of English. Word **2**, 171–187
Nishinuma, Y. (1979): Un modèle d'analyse automatique de la prosodie. Trav. Inst. Phon. Aix **6**
Nooteboom, S. G. (1980): "Speaking and Unspeaking: Detection and Correction of Phonological and Lexical Errors in Spontaneous Speech", in *Errors in Linguistic Performance,* ed. by V. A. Fromkin (Academic, New York) pp. 87–95

O'Connell, D. C., Kowal, S. (1972): Cross-linguistic pause and rate phenomena in adults and adolescents. J. Psycholinguist. Res. **1**, 155–165
O'Connell, D. C., Kowal, S., Hörmann, H. (1969): Semantic determinants of pauses. Psychol. Forsch. **33**, 50–67
O'Connor, J. D., Arnold, G. F. (1961): *Intonation of Colloquial English* (Longmans, London)
Öhman, S. E. G. (1967): "Word and Sentence Intonation: A Quantitative Model", Speech Transmission Laboratory (Stockholm) Quarterly Progress and Status Report, 2–1967, pp. 20–54
Ohala, J. (1975): Review of Lehiste 1970. Language **51**, 736–740
Ohala, J. (1978): "Production of Tone", in *Tone: A Linguistic Survey,* ed. by V. A. Fromkin (Academic, New York)
Ohala, J., Ewan, W. (1973): Speed of pitch range. J. Acoust. Soc. Am. **53**, 345
Oller, D. K. (1973): The effect of position in utterance on speech segment duration in English. J. Acoust. Soc. Am. **54**, 1235–1247
Oller, D. K., Smith, B. L. (1977): Effect of final-syllable position on vowel duration in infant babbling. J. Acoust. Soc. Am. **62**, 944–997
O'Malley, M. H., Kloker, D. R., Dara-Abrams, B. (1973): Recovering parentheses from spoken algebraic expressions. IEEE Trans. Audio Electroacoust. **AU-21**, 217–220
O'Shaughnessy, D. (1976): "Modelling Fundamental Frequency and Its Relationship to Syntax, Semantics and Phonetics", Ph. D. Thesis, Massachussetts Institute of Technology
Pellowe, J., Jones, V. (1979): Establishing intonationally variable systems in a multi-dimensional linguistic space. Lang. Speech **22**, 97–116
Perkell, J. S. (1980): "Phonetic Features and the Physiology of Speech Production", in *Language Production. Volume I: Speech and Talk,* ed. by B. Butterworth (Academic, London) pp. 337–372
Pierrehumbert, J. B. (1979): The perception of fundamental frequency declination. J. Acoust. Soc. Am. **66**, 363–369
Pierrehumbert, J. B. (1980): "The Phonology and Phonetics of English Intonation", Ph. D. Thesis, Massachussetts Institute of Technology
Pierrehumbert, J. B. (1981): Synthesizing intonation. J. Acoust. Soc. Am. **70**, 985–995
Pike, K. L. (1945): *The Intonation of American English* (University of Michigan Press, Ann Arbor)
Prince, E. F. (1981): "Toward a Taxonomy of Given/New Information", in *Radical Pragmatics,* ed. by P. Cole (Academic, New York) pp. 223–255
Rapp, K. (1971): "A Study of Syllable Timing", Speech Transmission Laboratory (Stockholm), Quarterly Progress and Status Report, 1–1971, pp. 14–19
Rischel, J. (1960): Über die phonematische and morphophonematische Funktion der sogenannten Worttöne im Norwegischen. Z. Phonet. Sprachwiss. Kommunikationsforsch. **13**, 177–185
Rischel, J. (1972): Compound stress in Danish without a cycle. Annu. Rep. Inst. Phon., Univ. Copenhagen **6**, 211–228
Rischel, J. (1975): "Problemer ved en generativ beskrivelse af dansk tryk", Selskab for Nordisk Filologi, Årsberetning for 1971–73, pp. 22–32
Rischel, J. (1980): Phrasenakzent als Signal des Objekts ohne "determiner" im Dänischen. Kopenhagener Beitr. german. Linguist., Suppl. 1, 262–279
Rischel, J. (in press): On unit accentuation – and the distinction between deep and surface phonology. Folia Linguist. **17**
Salles, J. L. (1980): "Essai d'un Enseignement de l'Intonation aux Jeunes Déficients Auditifs Profonds par Visualisation Graphique", Thèse, Centre de l'Ouïe et de la Parole, Fougères, France
Sanford, A. J., Garrod, S. C. (1981): *Understanding Written Language* (Wiley, London)
Schane, S. (1972): "Non-cyclic Word Stress", in *Essays on the Sound Pattern of English,* ed. by D. Goyvaerts, G. Pullum, pp. 249–259
Scherer, K. R. (1981): "Speech and Emotional States", in *The Evaluation of Speech in Psychiatry and Medicine,* ed. by J. Darby (Grune and Stratton, New York)

Scherer, U., Scherer, K. R. (1980): "Psychological Factors in Bureaucratic Encounters: Determinants and Effects of Interactions Between Officials and Clients", in *The Analysis of Social Skills,* ed. by W. T. Singleton, P. Spurgeon, R. B. Stammers (Plenum, New York) pp. 315–328

Scott, D. R. (1982): Duration as a cue to the perception of a phrase boundary. J. Acoust. Soc. Am. **71**, 996–1007

Selkirk, E. O. (1972): "The Phrase Phonology of English and French", Ph. D. Thesis, Massachusetts Institute of Technology

Selkirk, E. O. (1978a): On the status of "mute" e. Stud. Fr. Linguist. **1**, 141–150

Selkirk, E. O. (1978b): "On Prosodic Structure and Its Relation to Syntactic Structure", Conference on the Mental Representation of Phonology, Amherst, MA (Distributed by Indiana University Linguistics Club)

Selkirk, E. O. (1980): The role of prosodic categories in English word stress. Linguist. Inq. **11**, 563–605

Shields, J. L., McHugh, A., Martin, J. G. (1974): Reaction time to phoneme targets as a function of rhythmic cues in continuous speech. J. Exp. Psychol. **102**, 250–255

Sorin, C. (1981): Functions, roles and treatments of intensity in speech. J. Phon. **9**, 359–374

Stewart, J. M. (1971): "Niger-Congo, Kwa", in *Current Trends in Linguistics, Vol. VII: Linguistics in Sub-Saharan Africa,* ed. by T. A. Sebeok (Mouton, The Hague) pp. 179–212

Stock, E. (1980): *Untersuchungen zu Form, Bedeutung und Funktion der Intonation im Deutschen* (Akademie, Berlin)

Stockwell, R. P. (1972): "The Role of Intonation: Reconsiderations and Other Considerations", in *Intonation,* ed. by D. L. Bolinger (Penguin, Harmondsworth) pp. 87–109

Stowell, T. (1979): "Stress-Systems of the World Unite", in Papers on Syllable Structure, Metrical Structure and Harmony Processes, MIT Working Papers on Linguistics, Vol. 1, ed. by K. Safir, pp. 51–76

Svetozarova, N. D. (1975): "The Inner Structure of Intonation Contours in Russian", in *Automatic Analysis and Perception of Speech,* ed. by G. Fant, M. A. A. Tatham (Academic, London)

Takefuta, Y. (1975): "Method of Acoustic Analysis of Intonation", in *Measurement Procedures in Speech, Hearing and Language,* ed. by S. Singh (University Park Press, Baltimore) pp. 368–378

Terken, J. M. B. (1980a): "The Relation Between Information Structure and Accentuation: Some Exploratory Investigations", Institute for Perception Research Report 381, pp. 1–40

Terken, J. M. B. (1980b): "The Distribution of Pitch Accent in Descriptive Language as a Function of Informational Variables", IPO Ann. Prog. Rep. **15**, pp. 48–53

't Hart, J. (1979): Explorations in automatic stylization of F_0 curves. IPO Annu. Prog. Rep. **14**, 61–65

't Hart, J., Cohen, A. (1973): Intonation by rule: A perceptual quest. J. Phon. **1**, 309–327

't Hart, J., Collier, R. (1975): Integrating different levels of intonation analysis. J. Phon. **3**, 235–255

't Hart, J., Collier, R. (1978): A course in Dutch Intonation. IPO Annu. Prog. Rep. **13**, 31–35

't Hart, J., Collier, R. (1979): "On the Interaction of Accentuation and Intonation in Dutch", Proc. 9th Int. Congr. Phonetic Sciences, Copenhagen, Vol. 2, pp. 395–402

Thompson, H. S. (1980): *Stress and Salience in English: Theory and Practice*, (Xerox Corporation, Palo Alto)

Thorsen, N. (1978): An acoustical analysis of Danish intonation. J. Phon. **6**, 151–175

Thorsen, N. (1979): Interpreting raw fundamental frequency tracings of Danish. Phonetica **36**, 57–58

Thorsen, N. (1980a): Neutral stress, emphatic stress, and sentence intonation in Advanced Standard Copenhagen Danish. Annu. Rep. Inst. Phon., Univ. Copenhagen **14**, 121–205

Thorsen, N. (1980b): A study of the perception of sentence intonation – Evidence from Danish. J. Acoust. Soc. Am. **67**, 1014–1030

Thorsen, N. (1980c): Intonation contours and stress group patterns in declarative sentences of varying length in ASC Danish. Annu. Rep. Inst. Phon., Univ. Copenhagen **14**, 1–29

Thorsen, N. (in press): Standard Danish sentence intonation – Phonetic data and their representation. Folia Linguist. **17**

Touati, P. (unpublished): "Etudes Contrastives de la Prosodie du Français et du Suédois"

Trager, G. L., Smith, H. L. (1951): *An Outline of English Structure* (Battenburg, Norman, OK)

Umeda, N., Quinn, A. M. S. (1981): Word duration as an acoustic measure of boundary perception. J. Phon. **9**, 19–28

Vaissière, J. (1971): "Contribution à la Synthèse par Règles du Français", Thèse de Troisième cycle, Grenoble, France

Vaissière, J. (1974): "On French Prosody", Research Laboratory of Electronics (MIT), Quarterly Progress Report **114**, pp. 212–223

Vaissière, J. (1975): "Further Note on French Prosody", Research Laboratory of Electronics (MIT), Quarterly Progress Report **115**, pp. 251–262

Vaissière, J. (1980): La structuration acoustique de la phrase française. Ann. Scu. Norm. Sup. Pisa, Ser. III, **10**, 529–560

Van Lancker, D., Fromkin, V. A. (1973): Hemispheric specialization for pitch and "tone": Evidence from Thai. J. Phon. **1**, 101–109

Venneman, T. (1972): On the theory of syllabic phonology. Linguist. Ber. **18**, 1–18

Venneman, T. (1978): Universal syllabic phonology. Theor. Linguist. **5**, 175–215

Venneman, T. (1979): "An Outline of Universal Phonology", unpublished manuscript, University of Munich

Vergnaud, J.-R., Halle, M. (1978): "Metrical Phonology: A Fragment of a Draft", unpublished manuscript, Massachusetts Institute of Technology (revised 1979)

von Essen, O. (1956): Hochdeutsche Satzmelodie. Z. Phonet. Sprachwiss. Kommunikationsforsch. **9**, 75–85

Wang, W. S-Y (1967): Phonological features of tone. Int. J. Am. Linguist. **33**, 93–105

Welmers, W. E. (1959): Tonemics, morphotonemics and tonal morphemes. Gen. Linguist. **4**, 1–9

Weniger, D. (1978): "Zur Prosodie bei Aphasie", in *Brennpunkte der Patholinguistik,* ed. by G. Peuser (Fink, Munich) pp. 305–325

Wheeler, D. (1981): "Aspects of a Categorial Theory of Phonology", Ph. D. Thesis, University of Massachusetts

Williams, C. E., Stevens, K. N. (1972): Emotions and speech: Some acoustical correlates. J. Acoust. Soc. Am. **52**, 1238–1250

Yule, G. (1980): Intonation and givenness in spoken discourse. Stud. Lang. **4**, 271–286

Yule, G. (1981 a): "Aspects of the Information Structure of Spoken Discouse", Ph. D. Thesis, University of Edinburgh

Yule, G. (1981 b): New, current and displaced entity reference. Lingua **55**, 41–52

Zee, E., Maddieson, I. (1979): Tones and tone sandhi in Shanghai: Phonetic evidence and phonological analysis. UCLA Work. Pap. Phon. **45**, 93–129

Zingle, H. (1974): Quelques remarques sur le groupe accentuel en allemand. Trav. Inst. Phon., Strasbourg **6**, 89–117

Zlatoustova, L. V. (1954): Duration of vowel and consonant sounds in the Russian language. Scientific Notes, Kazan State University **114**, 99 (in Russian)

Zurif, E. B., Mendelsohn, M. (1972): Hemispheric specialization for the perception of speech sounds: The influence of intonation and structure. Percept. Psychophys. **11**, 329–332

Name Index

Names of languages are listed in italics

W. E. Cooper, J. M. Sorensen

Fundamental Frequency in Sentence Production

1981. 85 figures. X, 213 pages
ISBN 3-540-90510-3

Contents: Introduction. - Declination. - Fall-Rise Patterns. - Blocking. - Conclusions. - Extensions. - References. - Author Index. - Subject Index.

The Production of Speech

Editor: **P. F. MacNeilage**
1983. 90 figures. XVI, 302 pages
ISBN 3-540-90735-1

Contents: Control Processes in Speech Production: Central Processes Involved in Arm Movement Control. Laryngeal Gestures in Speech Production. Dynamic Characteristics of Voice Fundamental Frequency in Speech and Singing. The Segmental Organization of Speech. Development of Speech Production: Perspectives from Natural and Perturbed Speech. Sublexical Units and Suprasegmental Structure in Speech Production Planning. A "Dynamic Pattern" Perspective on the Control and Coordination of Movement. - Production Constraints and Sound Patterns of Language: Cross-Linguistic Studies of Speech Production. The Origin of Sound Patterns in Vocal Tract Constraints. Economy of Speech Gestures. Design Features of Speech Sound Systems. - Two Perspectives: In Favor of Some Uncommon Approaches to the Study of Speech. Some Reflections on Speech Research. - Author Index. - Subject Index. - Language Index.

A. C. Graesser

Prose Comprehension Beyond the Word

1981. 27 figures. X, 310 pages
ISBN 3-540-90544-8

Contents: A Schema-Based Framework for Representing Knowledge. - Allocation of Cognitive Resources During Prose Comprehension. - A Schema Pointer Plus Tag Model of Prose Memory. - The Conceptual Representation of Prose in Memory. - The Utilization of Prose Representations. - The Construction of Prose Representations. - References. - Author Index. - Subject Index.

**Springer-Verlag
Berlin Heidelberg
New York Tokyo**

H. Hörman

To Mean - To Understand

Problems of Psychological Semantics
Translated from the German by B. A. Jankowski
1981. 5 figures. X, 337 pages
ISBN 3-540-10448-8

Contents: The Problem of Signs. - The Rise of Psycholinguistics. - Competence and Performance. - Grammar: Syntax Plus a Poor Relative. - The Lexicon and Its Dimensions. - Functional Analyses of Semantic Dimensions. - The Concept of Sense Constancy. - Generative Semantics. - Philosophy of Language as Aide and Ally. - Russian Psychology of Language. - The Nonverbal Setting of Language. - The Genesis of Language. - Utterances Are Ego-Centered. - What Goes On in a Sentence. - Levels and Vectors of Understanding. - Conclusions. - References. - Author Index. - Subject Index.

H. Hörmann

Psycholinguistics

An Introduction to Research and Theory
Translated from the German by H. H. Stern, P. Leppmann
2nd revised edition. 1979. 60 figures, 21 tables.
IX, 342 pages. ISBN 3-540-90417-4

Contents: Introduction: Organism, Language, and World. - Sign, Expression, and Symbol Linguistic Units and the Rules for Their Connection. - Language, Information, and Communication. - Frequency and Probability. - Verbal Association and the Problem of Meaning. - The Philosophical Background to Modern Psychology of Language. - Sign and Object Signified: Classical Theories of the Development of Meaning. - Imitation of Sound, Sound Symbolism, Expression. - The Psychological Reality of Grammar. - The Developmental Psychology of Language Acquisition. - Language Comprehension and Man's View of the World. - Bibliography. - Author Index. - Subject Index.

Hearing Mechanisms and Speech

EBBS-Workshop, Göttingen, April 26–28, 1979
Editors: **O. Creutzfeldt, H. Scheich, C. Schreiner**
1979. 85 figures, 12 tables. XXIII, 413 pages
ISBN 3-540-09655-8

Contents: Functional and Structural Conditions for Transmission of Auditory Signals in the Nervous System. - The Representation of Complex Natural Sounds in the Auditory System. - Brain Mechanisms of Speech Perception and Production. - Psychoacoustic Elements in Language. - Linguistic Elements in Speech Perception. - Receptive Defects of Linguistic Elements in Aphasia.

Springer Series in Language and Communication

Editor: **W. J. M. Levelt**

Volume 1
W. Klein, N. Dittmar
Developing Grammars
The Acquisition of German Syntax by Foreign Workers
1979. 9 figures, 38 tables. X, 222 Seiten
ISBN 3-540-09580-2

Volume 2
The Child's Conception of Language
Editors: **A. Sinclair, R. J. Jarvella, W. J. M. Levelt**
1978. 9 figures, 5 tables. IX, 268 pages
ISBN 3-540-09153-X

Volume 3
M. Miller
The Logic of Language Development in Early Childhood
Translated from the German By R. T. King
1979. 1 figure, 30 tables. XVI, 478 pages
ISBN 3-540-09606-X

Volume 4
L. G. M. Noordman
Inferring from Language
With a Foreword by H. H. Clark
1979. 4 figures, 25 tables. XII, 170 pages
ISBN 3-540-09386-9

Volume 5
W. Noordman-Vonk
Retrieval from Semantic Memory
With a Foreword by J. C. Marshall
1979. 10 figures, 19 tables. XII, 97 pages
ISBN 3-540-09219-6

Volume 6
Semantics from Different Points of View
Editors: **R. Bäuerle, U. Egli, A. v. Stechow**
1979. 15 figures, 7 tables. VIII, 419 pages
ISBN 3-540-09676-0

Volume 7
C. E. Osgood
Lectures on Language Performance
1980. 31 figures, 33 tables. XI, 276 pages
ISBN 3-540-09901-8

Volume 8
T. Ballmer, W. Brennenstuhl
Speech Act Classification
A Study in the Lexical Analysis of English Speech Activity Verbs
1981. 4 figures. X, 274 pages
ISBN 3-540-10294-9

Volume 9
D. T. Hakes
The Development of Metalinguistic Abilities in Children
In collaboration with J. S. Evans and W. Tunmer
1980. 6 figures, 8 tables. X, 119 pages
ISBN 3-540-10295-7

Volume 10
R. Narasimhan
Modelling Language Behavior
1981. 3 figures. XVI, 217 pages
ISBN 3-540-10513-1

Springer-Verlag
Berlin
Heidelberg
New York
Tokyo

MIX
Papier aus verantwortungsvollen Quellen
Paper from responsible sources
FSC® C105338

If you have any concerns about our products,
you can contact us on
ProductSafety@springernature.com

In case Publisher is established outside the EU,
the EU authorized representative is:
Springer Nature Customer Service Center GmbH
Europaplatz 3, 69115 Heidelberg, Germany

Printed by Libri Plureos GmbH
in Hamburg, Germany